Lecture Notes in Mathematics

Edited by A. Dold and B. Ec'

T0216765

1363

Peter G. Casazza
Thaddeus J. Shura

Tsirelson's Space

With an Appendix by J. Baker, O. Slotterbeck and R. Aron

Springer-Verlag

Berlin Heidelberg New York London Paris Tokyo

Authors

Peter G. Casazza
Department of Mathematics, University of Missouri
Columbia, MO 65211, USA

Thaddeus J. Shura
Kent State University, Salem Campus
South Salem OH 44460, USA

Mathematics Subject Classification (1980): 46 B 20, 46 B 25

ISBN 3-540-50678-0 Springer-Verlag Berlin Heidelberg New York
ISBN 0-387-50678-0 Springer-Verlag New York Berlin Heidelberg

Printing and binding: Druckhaus Beltz, Hemsbach/Bergstr.
2146/3140-543210

Dedication

We dedicate these notes to Professor R. C. James.

Preface

One of the historical concerns of the structure theory of Banach spaces has been whether there were any 'fundamental" spaces which embedded isomorphically in every infinite-dimensional Banach space. ". . . from the point of view of the theory of classical Banach spaces the 'nicest' subspace one could possibly hope to find in a general Banach space would be either c_0 or $\ell_p(1 \leq p < \infty)$. The feeling that this could be the case was based on the fact that all classical spaces do indeed contain a copy of c_0 or $\ell_P(1 \leq p < \infty)$. Also Orlicz spaces have this property despite the fact that . . . the definition of an Orlicz space is not a priori connected to any ℓ_p space or c_0." [55]

The classical hope that c_0 or some ℓ_p always embeds in a general Banach space was fueled by some strong results hinting at how very important these spaces are. We list only a few here:

1. A Banach space X contains an isomorph of c_0 if and only if there is a sequence $\{x_n\}_{n=1}^{\infty}$ in X such that $\sum_n |x^*(x_n)| < \infty$, $\forall x^* \in X^*$, but $\sum_n x_n$ does not converge.

2. Every bounded sequence in a Banach space has a subsequence which is either weakly Cauchy or equivalent to the unit vector basis of ℓ_1.

3. Let X be a Banach space with a normalized basis with the property that this basis is equivalent to each of its normalized block bases. Then this basis is already equivalent to the unit vector basis of c_0 or some $\ell_p(1 \leq p < \infty)$.

4. If an infinite-dimensional Banach space has an unconditional basis, then it contains uniformly complemented copies of ℓ_1^n, ℓ_2^n, or ℓ_∞^n, $\forall n$.

That this hope was unfounded was demonstrated by a (surprisingly simple) construction of B. S. Tsirelson of a reflexive Banach space with monotone unconditional Schauder basis and no embedded copies of c_0 or any ℓ_p. His example opened a Pandora's box of pathological variations, and has had a tremendous effect upon the study of Banach spaces. The original construction has been "dualized" [26], "treed" [51], "symmetrized" [26], "convexified" [26], and "modified" [44]. A "generalized" Tsirelson's space, a related space of holomorphic functions [1], and a computer program which efficiently calculates the norm [appendix] were all out-growths of the original example.

The basic properties of this space were developed in [16], [17], and [26], where it was shown that Tsirelson's space provided a single Banach space with a complex array of properties against which mathematicians could sharpen (or nullify) many conjectures. The space itself became an object of study: it was shown to have the fixed point property in [25], its spreading models have been studied in [12], and its complemented subspaces have been studied (this latter topic forms one of our chapters). Recently its isometries have been characterized (see chapter III).

Our objectives in preparing this manuscript were three-fold:

a) to collect what is known about the space,

b) to organize and unify this into a coherent theory,

c) and to point the way towards further research.

Although we have tried to make the basic construction accessible to as wide an audience as possible, we have not been reluctant to include "fresh" results which may indeed become polished with the passage of time. Roughly a third of this material consists of previously unpublished results.

The background required for reading these notes is a knowledge of the theory of Schauder bases and of classical Banach spaces. The excellent books Classical Banach Spaces I, II [35,36] make good collateral reading. Most chapters end with a section of notes and remarks which frequently contain variant constructions, additional theory (of an ancilliary nature), open questions, and kudos. Definitions are supplied when felt necessary.

It is our belief that one of the goals of functional analysis ought to be the study of individual spaces, while it is our hope that such a study will eventually enrich the entire field. We have tried to make the basic construction accessible to graduate students with a course in functional analysis (and some basis theory) under their belts. For that reason some of the proofs in the first four chapters may seem inefficient to the specialist. Chapter 0 can be read independently of the other chapters, while chapters I-III contain basic structural results which are leaned upon throughout all that follows.

We thank B. Beauzamy for some helpful suggestions, T. Barton for his reading and recommendations, the Department of Mathematical Sciences at Kent State University for the use of its VAX 11-780, Gail E. Bostic, Ava D. Logsdon and Shirley M. Sommers for their heroic efforts at computerized type-production, and Janet Tremain (the original JT*!) for her support. The second writer also thanks Dean J. Cooney and the Regional Campus System of Kent State University for their support.

TABLE OF CONTENTS

Precursors of the Tsirelson construction.

The recursive construction which we present in the next chapter appears (at first sight) very unnatural. In order to ease our way into the construction of Tsirelson's space T, we present here some conceptual predecessors of the Figiel-Johnson rendering of Tsirelson's space. These less complicated kindred constructions will (hopefully) facilitate the introduction of T and help place it into a proper historical perspective. We begin by drawing upon some notions that have been around since the days of Stefan Banach himself.

In 1930, J. Schreier [52] introduced the notion of "admissibility" while producing a counter-example to a question of S. Banach and S. Saks. These two had just shown [9] that in the spaces $L_p[0,1]$ (where $p > 1$) each weakly convergent sequence contains a subsequence whose arithmetic means converge in norm. They went on to ask whether such a thing held in the space of continuous functions $C[0,1]$. J. Schreier showed that this was not the case. A slight variation in his original concept produces the space which we (and folklore) call "Schreier's space S".

Definition 0.1: A finite subset $E = \{n_1 < n_2 < n_3 \cdots < n_k\}$ of natural numbers is said to be admissible if $k \leq n_1$. We denote by N the class of all admissible subsets of the natural numbers $\mathbb{N}$.

Construction 0.2: Let $\mathbb{R}^{(\mathbb{N})}$ denote the (vector) space of all real sequences which are eventually 0 (i.e., have "finite support"). Schreier's space S is the $\|\cdot\|_S$-completion of $\mathbb{R}^{(\mathbb{N})}$, where $\|x\|_S = \sup_{E \in N} \sum_{k \in E} \|x_k\|$, where x_k is the k-th coordinate of x, and $x \in \mathbb{R}^{(\mathbb{N})}$.

It's an exercise to see that $\|\cdot\|_S$ norms $\mathbb{R}^{(\mathbb{N})}$. We depart from the usual notation for the canonical unit vector basis by denoting by $\{s_n\}_{n=1}^\infty$ that sequence in S defined by:

$$s_n(k) = \begin{cases} 1, & \text{if } n = k \\ 0, & \text{otherwise.} \end{cases}$$

Recall

Definition 0.3: A basic sequence (basis) $\{x_n\}_{n=1}^\infty$ is 1-unconditional ("monotone unconditional") if $\|\sum_j a_j x_j\| \leq \|\sum_j b_j x_j\|$, for all scalar sequences $\|a_j\| \leq \|b_j\|, (1 \leq j < \infty)$.

Proposition 0.4: In Schreier's space S, we have:

(i). for $x = \sum_k x_k s_k \in \mathbb{R}^{(\mathbb{N})}, \|x\|_S \leq \|x\|_{\ell_1}$.

(ii) $\|s_n\|_S = 1, \forall n$.

(iii) $\{s_n\}_{n=1}^\infty$ is a 1-unconditional basis for S.

(iv) for $x \in \mathbb{R}^{(\mathbb{N})}$ such that supp $(x) \in N, \|x\|_S = \|x\|_{\ell_1}$. (For $x = \sum_k x_k s_k$ we define the support of x, supp $(x) = \{k : x_k \neq 0\}$.)

(v) the canonical injection from ℓ_1 into S has norm 1.

1

More importantly, we have the following:

Theorem 0.5: ℓ_1 does not embed into S.

Proof: If ℓ_1 did embed into S, by a theorem of R. C. James, it would do so nicely–i.e., we could find a sequence of normalized consecutive blocks $\{u_n\}_{n=0}^\infty$ against the basis $\{s_n\}_{n=1}^\infty$ for which:

$$\|\sum_i a_i u_i\|_S \geq \frac{9}{10}\sum_i |a_i|,$$

for every finite sequence of scalars $\{a_i\}_{i=0}^\infty$.

But then $\|u_0 + \frac{1}{n}(u_1 + \cdots + u_n)\| \geq \frac{18}{10}$, for all $n \geq 1$, in violation of the following Proposition.
$\square$

Proposition 0.6: For every sequence of normalized consecutive blocks $\{u_n\}_{n=0}^\infty$ against the basis $\{s_n\}_{n=1}^\infty$, we have:

$$\overline{\lim_n} \|u_0 + \frac{1}{n}(u_1 + \cdots + u_n)\|_S \leq 1.$$

Proof: For $n \geq 1$, consider the quantity $\|u_0 + \frac{1}{n}(u_1 + \cdots + u_n)\|_S$, and let $E \in N$ be such that this equals $\sum_{k\in E} |u_0(k) + \frac{1}{n}(u_1(k) + \cdots + u_n(k))|$.

If E does not meet supp (u_0), this second quantity is at most 1.

If E does meet supp (u_0), $\overline{\overline{E}}$ $(:=$ the cardinality of $E)$ is independent of n, and thus (by a triangulation) $\lim_n \sum_{k\in E} |\frac{1}{n}(u_1(k) + \cdots + u_n(k))| = 0$, and we're done. $\square$

We have as a companion result:

Proposition 0.7: c_0 embeds (isometrically) into S.

Proof:

We produce a sequence $\{u_n\}_{n=1}^\infty$ in S equivalent to the (canonical) unit vector basis of c_0, by defining:

$$
\begin{aligned}
u_1 &= s_1,\\
u_2 &= \tfrac{1}{2}(s_2 + s_3),\\
u_3 &= 2^{-2}(s_4 + s_5 + s_6 + s_7),\\
\end{aligned}
$$

.

.

.

$$u_n = 2^{1-n}\left(\sum_{k=2^{n-1}}^{2^n-1} s_k\right), (n \geq 3).$$

These u_n's are normalized consecutive blocks against the s_n's, each having admissible support. Letting $\{a_i\}_{i=1}^K$ be any scalars, we thus have:

$$\|\sum_{i=1}^K a_i u_i\|_S \geq |a_n|, \text{ for } 1 \leq n \leq K.$$

2

Thus, $\|\sum_{i=1}^{K} a_i u_i\|_S \geq \max_{n=1}^{K} |a_n|$.

By Proposition 0.4,

$$\|\sum_{i=1}^{K} a_i u_i\|_S = \|\sum_{i=1}^{K} |a_i| u_i\|_S \leq$$

$$\|\sum_{i=1}^{K} \left(\max_{k=1}^{K} |a_k| \right) u_i\|_S = \left(\max_{k=1}^{K} |a_k| \right) \|\sum_{i=1}^{K} u_i\|_S,$$

and it suffices now to show that

$$\|\sum_{i=1}^{K} u_i\|_S = 1.$$

Toward this end, let $\dot{u}_K = \sum_{i=1}^{K} u_i$, and choose $E \in N$ such that

$$\|\dot{u}_K\|_S = \sum_{j \in E} |\dot{u}_K(j)|.$$

Without loss of generality, E may be taken as a segment of natural numbers, say

$$E = \{\ell, \ell + 1, \ell + 2, \cdots, 2\ell - 1\},$$

since if $\dot{u}_K$ is written as

$$\dot{u}_K = \sum_{k=1}^{2^K - 1} a_k s_k,$$

a_k decreases through values $1, \frac{1}{2}, \frac{1}{4}$, etc.

Thus $\exists k_0$ such that

$$s^{k_0} \leq \ell < 2^{k_0+1}, \text{ and}$$

$$\|\dot{u}_K\|_S = \sum_{j \in E} |\dot{u}_K(j)| = 2^{-k_0} \left(2^{k_0+1} - \ell \right) + 2^{-k_0-1} \left(2\ell - 2^{k_0+1} \right) = 1.$$

$\square$

¿From Proposition 0.7, we obtain:

Corollary 0.8: Schreier's space S is not reflexive.

Proof: It's a well-known result of R. C. James that an infinite-dimensional Banach space with unconditional basis is reflexive iff it contains no isomorphic copies of ℓ_1 or c_0. $\square$

In 1972, A. Baernstein [8] produced a reflexive variant of S (we abuse chronology a bit here) which contains a weakly null sequence such that no subsequence is strongly Cesaro convergent. Baernstein's space B comes a bit closer to the flavor of the construction which we will adopt for Tsirelson's space T. The notations and usages introduced here will be of use in T and its myriad of variations.

Construction 0.9:

a) If E and F are finite non-void subsets of $\mathbb{N}$, we write "$E < F$" for

$$\max E < \min F,$$

(with similar meanings attached to "$E \leq F$", etc.). We write "$n < E$", instead of "$\{n\} < E$."

b) For $x \in \mathbb{R}^{(\mathbb{N})}$, we write Ex to indicate that vector defined by:

$$(Ex)(k) = \begin{cases} x(k), & \text{if } k \in E, \\ 0, & \text{otherwise.} \end{cases}$$

c) For $x \in \mathbb{R}^{(\mathbb{N})}$, we define $\|x\|_B =$

$$\sup\left\{\left(\sum_{k=1}^{n} \|E_k x\|_{\ell_1}^2\right)^{\frac{1}{2}} : E_k \in N, \text{ and } E_1 < E_2 < \cdots < E_n, \ n = 1, 2, \cdots\right\},$$

i.e., the "sup" is taken over all strings of consecutive finite subsets E_k of $\mathbb{N}$ such that $\overline{\overline{E}}_k \leq E_k$.

d) Baernstein's space B is the $\|\cdot\|_B$-completion of $\mathbb{R}^{(\mathbb{N})}$. (It's easily seen that $\|\cdot\|_B$ norms $\mathbb{R}^{(\mathbb{N})}$.)

If we denote the canonical unit vector basis of B by $\{b_n\}_{n=1}^{\infty}$, the following facts are easily entailed from definitions:

Proposition 0.10: In Baernstein's space B,

(i) $\{b_n\}_{n=1}^{\infty}$ is a monotone unconditional basis.

(ii) $\|b_n\|_B = 1$, for all n

More importantly, we have:

Theorem 0.11:

a) ℓ_1 does not embed in B.

b) c_0 does not embed in B.

c) B is reflexive.

Proof: (c): follows from (a), (b), Proposition 0.10, and the theorem of R. C. James mentioned earlier.

(b): By another result of R.C. James [59], it suffices to show that $\{b_n\}_{n=1}^{\infty}$ is boundedly complete, i.e., we need to show that for any scalar sequence $\{x(i)\}_{i=1}^{\infty}$, if $\sup_n \|\sum_{i=1}^{n} x(i)b_i\|_B \leq 1$, then $x \in B$. Toward this end, let $\{E_k\}_{k=1}^{\infty}$ be a family of consecutive admissible subsets of $\mathbb{N}$.

For each $K \in \mathbb{N}$,

$$\left(\sum_{k=1}^{K} \|E_k x\|_{\ell_1}^2\right)^{\frac{1}{2}} \leq \left\| \left(\bigcup_{k=1}^{K} E_k\right) x \right\|_B \leq 1.$$

Thus $\left(\sum_{k=1}^{\infty} \|E_k x\|_{\ell_1}^2\right)^{\frac{1}{2}} \leq 1$, and $x \in B$.

Thus we're done if we demonstrate (a), which follows from the next Proposition. $\square$

Proposition 0.12: For every sequence of normalized consecutive blocks $\{u_n\}_{n=0}^{\infty}$ against the basis $\{b_n\}_{n=1}^{\infty}$, we have:

$$\overline{\lim_n} \left\| u_0 + \frac{1}{n}(u_1 + u_2 + \cdots + u_n) \right\|_B \leq \sqrt{2}.$$

Proof: Fix $n \geq 1$, and choose a sequence of consecutive admissible subsets of $\mathbb{N}$, say $E_1 < E_2 < \cdots$.

If $\mathrm{supp}\,(u_0) < E_1$, then

$$\left(\sum_{k=1}^{\infty} \left\| E_k \left(u_0 + \frac{1}{n}(u_1 + \cdots + u_n)\right) \right\|_{\ell_1}^2\right)^{\frac{1}{2}} =$$

$$\left(\sum_{k=1}^{\infty} \left\| E_k \left(\frac{1}{n}(u_1 + \cdots + u_n)\right) \right\|_{\ell_1}^2\right)^{\frac{1}{2}} \leq 1.$$

Otherwise let k_1 denote the last index k for which

$$\min E_k \leq \max\,\mathrm{supp}\,(u_0), \text{ and let}$$

$$f(u_0) = \max\,\mathrm{supp}\,(u_0).$$

Then $\overline{\overline{E}}_{k_1} \leq f(u_0)$, and so

$$\sum_{k=1}^{k_1} \left\| E_k \left(u_0 + \frac{1}{n}(u_1 + \cdots + u_n)\right) \right\|_{\ell_1}^2$$
$$\leq \left(\|u_0\|_B + \left\| E_{k_1} \left(\frac{1}{n}(u_1 + \cdots + u_n)\right) \right\|_B \right)^2$$
$$\leq \left(1 + \frac{1}{n}\sum_{i=1}^{n} \|E_{k_1} u_i\|_B \right)^2$$
$$\leq \left(1 + \frac{1}{n}\cdot f(u_0)\right)^2.$$

Thus $\left(\sum_{k=1}^{\infty} \left\| E_k \left(u_0 + \frac{1}{n}(u_1 + \cdots + u_n)\right) \right\|_{\ell_1}^2\right)^{\frac{1}{2}} =$

$$\left\{\sum_{k=1}^{k_1} \left\| E_k \left(u_0 + \frac{1}{n}(u_1 + \cdots + u_n)\right) \right\|_{\ell_1}^2 + \sum_{k=k_1+1}^{\infty} \left\| E_k \left(\frac{1}{n}(u_1 + \cdots + u_1)\right) \right\|_{\ell_1}^2\right\}^{\frac{1}{2}}$$

$$\leq \left\{\left(1 + \frac{1}{n}\cdot f(u_0)\right)^2 + 1\right\}^{\frac{1}{2}},$$

and this, for all n. We're done. $\square$

We've seen that Schreier's space S has no embedded copies of ℓ_1, though c_0 does embed in S. Baernstein's space B contains isomorphs of neither c_0 nor ℓ_1, while both spaces have unconditional bases. B, however, does contain an isomorphic copy of ℓ_2: we can construct a sequence of normalized consecutive blocks $\{u_n\}_{n=1}^\infty$ against the $\{b_n\}_{n=1}^\infty$ which is equivalent to the canonical unit vector basis of ℓ_2. (A variation of the argument in Proposition 0.7 does the trick.) As we will see shortly, Tsirelson's space T neither contains copies of c_0, nor of $\underline{\text{any}}$ $\ell_p(1 \leq p < \infty)$ and has a 1-unconditional basis. In fact, its pathology is much deeper than this.

Notes and Remarks:

1/. Much of the material in this chapter has been adapted from [12]. Spreading models of S, B, and T and some variations are also studied there, as are "hybrid" spaces such as Schreier-Orlicz space and Baernstein-Orlicz space.

2/. It is well known that every uniformly convex space has the Banach-Saks Property, and that every space with this property is reflexive. Baernstein's space B is reflexive without the Banach-Saks Property. Schreier's space S can be "interpolated" with ℓ_1 to produce another example of a reflexive space which fails the Banach-Saks Property. See [11, 12] for details and for a treatment of the Weak Banach-Saks Property and the Alternating Banach-Saks Property.

3/. Schreier's space S not only contains an isometric copy of c_0, but is "hereditarily c_0," i.e.:

Proposition 0.13: If X is any infinite-dimensional subspace of S, then c_0 embeds in X.

Proof. For such an $X, \exists$ a subspace Y isomorphic to the span of a normalized block basic sequence $\{u_n\}_{n=1}^\infty$, say $u_n = \sum\limits_{i=p_n+1}^{p_n+1} a_i s_i, (n = 1, 2, \ldots), (p_1 < p_2 < \cdots)$, where $\{s_i\}_{i=1}^\infty$ is the usual basis for S.

If for some $n, \sup\limits_{m \geq 0} \| \sum\limits_{i=n}^{m+n} u_i \|_S < \infty$, then (since $\{u_i\}_{i=n}^\infty$ is an unconditional basic sequence) $\{u_i\}_{i=n}^\infty$ is equivalent to the (canonical) unit vector basis of c_0.

If for all $n, \sup\limits_{m \geq 0} \| \sum\limits_{i=n}^{m+n} u_i \|_S = \infty$, we can select indices $n_1 < n_2 < \cdots$ such that:

$$\| \sum_{i=n_k+1}^{n_{k+1}} u_i \|_S \geq p_{n_k+1}, \quad (k = 1, 2, \cdots).$$

In this case, let

$$w_k = \frac{\sum\limits_{i=n_k+1}^{n_{k+1}} u_i}{\| \sum\limits_{i=n_k+1}^{n_{k+1}} u_i \|_S} = \sum_{i=p_{n_k+1}+1}^{p_{n_{k+1}+1}} \gamma_i S_i, \text{ say.}$$

We then have that:

1. $\|w_k\|_S = 1, \quad (k = 1, 2, \cdots)$, and

2. $|\gamma_i| \leq \frac{1}{p_{n_k+1}+1}$, for $p_{n_k+1} + 1 \leq i \leq p_{n_{k+1}+1})$.

(This latter must hold, since otherwise that u_j whose support contains i such that $|\gamma_i| > \frac{1}{p_{n_k+1}+1}$ would have norm > 1, yielding a contradiction.)

Thus $\|\sum_{k=1}^{m} w_k\|_S = \sup_{E \in N} \sum_{i \in E} |\gamma_i| = \sum_{i \in E_0} |\gamma_i|$, say, and if

$$p_{n_k+1} \leq \min E_0 \leq p_{n_k+1+1}, \text{ then}$$

$$\sum_{i \in E_0} |\gamma_i| \leq \|w_k\|_S + \frac{\bar{\bar{E}}_0}{p_{n_k+1}+1} \leq 2.$$

Hence $\sup_{m \geq 1} \|\sum_{k=1}^{m} w_k\|_S \leq 2$, and so $\{w_k\}_{k=1}^{\infty}$ is equivalent to the unit vector basis of c_0. $\square$

We note here that Proposition 0.13 provides an alternative proof for Theorem 0.5. We could argue as follows: S is c_0-rich, but ℓ_1 and c_0 are totally incomparable. We've included the given proof for Theorem 0.5 since it is nicely analogous to the corresponding result for Tsirelson's space.

4/. Baernstein's space B has some variants which have been studied in the thesis of C.J. Seifert [53], and are called (by him) "B_p" spaces.

Construction 0.14: Fix $1 < p < \infty$, and adapt Construction 0.9 by replacing the number 2 by p in (c). The resulting space is B_p. Seifert shows:

Theorem 0.15:

(a) For $1 < p < \infty$, B_p is reflexive, with unconditional basis, and lacks embedded isomorphs of c_0 and ℓ_1.

(b) B_p fails the Banach-Saks Property, although B_p^* has it.

(c) Every normalized block basic sequence against the canonical unit vector basis $\{b_n\}_{n=1}^{\infty}$ contains a subsequence which is equivalent to the canonical unit vector basis of ℓ_p. (In fact, the closed linear span of the subsequence is isometrically isomorphic to ℓ_p.)

(d) $\{b_{2n}\}_{n=1}^{\infty}(\subset B_p)$ is isometrically equivalent to the usual unit vector basis of ℓ_p.

(e) Let X be an infinite-dimensional subspace of B_p. Then X contains a subspace Y such that Y is isomorphic to ℓ^p and Y is complemented in B_p.

(f) Any bounded linear operator $L : B_p \to B_q$, (where $1 < q < p < \infty$), is compact.

(g) B_p and B_q, (where $1 < q < p < \infty$), are totally incomparable.

5/. The theorems of R.C. James which we use in this section are widely known. A good reference concerning them is [59].

Chapter I: The Figiel-Johnson construction of Tsirelson's space.

For Schreier's space S, we employed the notion of an "admissible" subset of the natural numbers, while Baernstein's space B required sequences of consecutive admissible subsets of $\mathbb{N}$. The construction of Tsirelson's space T which we use here is a bit more sophisticated than either of these, and is due to T. Figiel and W. B. Johnson [26]. In fact, it yields the dual of B. S. Tsirelson's original example [56] (which we discuss in the "Notes and Remarks" section at the end of this chapter). The Figiel-Johnson construction carries the advantage of giving an analytic (and computable) description of the norm on T, and it permits us to develop (inductively) certain sequential principles with which we can study operators, complemented subspaces, and subsequences of the canonical unit vector basis of T. The major properties of T known to B. S. Tsirelson are collected in Theorem I.8 (though we prove them individually). The material following Theorem I.8 is needed in almost every successive chapter, and largely carries over to the great many variations on T which infest these notes. Remarks are made concerning the extent of the "carry-over" wherever appropriate. (Indeed, if Theorem I.8 described the only interesting pathologies of T, we would not need quite so many pages!) We begin with:

Construction I.1:

(a) For E, F finite non-void subsets of $\mathbb{N}$, we write: "$E \leq F$", for "$\max E \leq \min F$", with "$n \leq E$", instead of "$\{n\} \leq E$", and with analogous meanings for "$E < F$", etc. (just as in Chapter 0).

(b) Let $\mathbb{R}^{(\mathbb{N})}$ be the (vector) space of all real scalar sequences of finite support. (Again, just as in Chapter 0.)

(c) Let $\{t_n\}_{n=1}^{\infty}$ denote the canonical unit vector basis of $\mathbb{R}^{(\mathbb{N})}$.

(d) For any $x = \sum_n a_n t_n \in \mathbb{R}^{(\mathbb{N})}$, and any $E \geq 1$, we define $Ex = \sum_{n \in E} a_n t_n$.

(e) We now inductively define a sequence of norms $\{\|\cdot\|_m\}_{m=0}^{\infty}$ upon $\mathbb{R}^{(\mathbb{N})}$ as follows:

$$\text{fixing } x = \sum_n a_n t_n \in \mathbb{R}^{(\mathbb{N})},$$

$$\text{let } \begin{cases} \|x\|_0 = \max_n |a_n|, \text{ and} \\ \|x\|_{m+1} = \max\left\{ \|x\|_m, \frac{1}{2}\max\left[\sum_{j=1}^{k} \|E_j x\|_m \right] \right\}, \text{(for } m \geq 0\text{)}, \end{cases}$$

where the "inner" max is taken over all choices of finite subsets $\{E_j\}_{j=1}^{k}$ of $\mathbb{N}$ as k varies and such that $k \leq E_1 < E_2 < \cdots < E_k$.

(Any such expression $\frac{1}{2}\sum_{j=1}^{k} \|E_j x\|_m$ is called an <u>admissible sum for x</u>.)

8

(f) It's easily seen that the $\| \cdot \|_m$ are norms on $\mathbb{R}^{(N)}$, that they increase with m, and that

$$\|x\|_m \leq \sum_n |a_n|, \text{ for all } x = \sum_n a_n t_n \in \mathbb{R}^{(N)},$$

and for all m. Thus, for each $x \in \mathbb{R}^{(N)}, \lim_m \|x\|_m$ exists and is majorized by $\|x\|_{\ell_1}$. We denote $\lim_m \|x\|_m$ by $\|x\|$, and easily confirm that it norms $\mathbb{R}^{(N)}$.

(g) Tsirelson's space T is the $\| \cdot \|$-completion of $\mathbb{R}^{(N)}$, where $\| \cdot \|$ is defined in (f), just above.

We will also denote by $\{t_n\}_{n=1}^{\infty}$ the canonical unit vector basis of T. Our construction of T differs slightly from that found in [26] and [35] by allowing $k \leq E_1$, instead of $k < E_1$. Later, it will become clear that the induced norms under either form of "admissibility" are equivalent. (We've chosen ours to simplify notation in a few complicated arguments.) Indeed, it is immediate that $\{t_n\}_{n=1}^{\infty}$ under our definition is isometrically equivalent to $\{t_n\}_{n=2}^{\infty}$ under the definition of [26] and [35].

The following facts are listed so that we can apply them in proofs of the non-embedding results which follow them. They all follow easily from the definitions involved.

Proposition I.2:

1. The sequence $\{t_n\}_{n=1}^{\infty}$ forms a normalized 1-unconditional Schauder basis for the space T.

2. For each $x = \sum_n a_n t_n \in T$,

$$\|x\| = \max \left\{ \max_n |a_n|, \frac{1}{2} \sup \sum_{j=1}^{k} \|E_j x\| \right\},$$

where the inner "sup" is taken over all choices

$$k \leq E_1 < E_2 < \cdots < E_k, \text{ and all } k.$$

3. For any $k \in \mathbb{N}$ and any k normalized blocks $\{y_i\}_{i=1}^{k}$, such that

$$y_i = \sum_{n=p_i+1}^{p_{i+1}} a_n t_n, \text{ with } \begin{cases} 1 \leq i \leq k, \text{ and} \\ k-1 \leq p_1 < p_2 < \cdots < p_{k+1}, \end{cases}$$

we have:

$$\frac{1}{2} \sum_{i=1}^{k} |b_i| \leq \left\| \sum_{i=1}^{k} b_i y_i \right\| \leq \sum_{i=1}^{k} |b_i|, \text{ for all scalars } \{b_i\}_{i=1}^{k}.$$

The following result and its proof should be compared to the corresponding results in Chapter 0 concerning Schreier's space and Baernstein's space.

Proposition I.3: ℓ_1 does not embed into T.

Proof. Suppose that it did. Then by a theorem of R. C. James [59, p. 257] it would have to do so nicely, i.e., there would exist a normalized block basis $\{y_i\}_{i=0}^{\infty}$ against $\{t_n\}_{n=1}^{\infty}$ such that for every choice of scalars $\{b_i\}_{i=0}^{\infty}$,

$$\frac{8}{9} \sum_{i=0}^{\infty} |b_i| \leq \left\| \sum_{i=0}^{\infty} b_i y_i \right\| \leq \sum_{i=0}^{\infty} |b_i|.$$

9

¿From this it follows that:

$$\|y_0 + \frac{1}{r}\sum_{i=1}^{r} y_i\| \geq \frac{16}{9}, \text{ for all } r = 1, 2, \cdots . \tag{1}$$

Now consider $k \leq E_1 < E_2 < \cdots < E_k$, and let $n_0 = \max \operatorname{supp} y_0$.

If $k > n_0$, then

$$\sum_{j=1}^{k} \|E_j \left(y_0 + \frac{1}{r}\sum_{i=1}^{r} y_i\right)\| = \sum_{j=1}^{k} \|E_j \left(\frac{1}{r}\sum_{i=1}^{r} y_i\right)\| \leq 2. \tag{2}$$

If $k \leq n_0$, we set

$$A = \{i : \|E_j y_i\| \neq 0 \text{ for at least two values of } j\}.$$
$$B = \{i : \|E_j y_i\| \neq 0 \text{ for at most one value of } j\}.$$

Then since A has at most k elements we have:

$$\sum_{j=1}^{k} \|E_j \left(y_0 + \frac{1}{r}\sum_{i=1}^{r} y_i\right)\| \leq \sum_{j=1}^{k} \|E_j y_0\| + \frac{1}{r}\left[\left(\sum_{i \in A}\sum_{j=1}^{k} + \sum_{i \in B}\sum_{j=1}^{k}\right)\|E_j y_i\|\right]$$
$$\leq 2\|y_0\| + \frac{1}{r}\left(2\sum_{i \in A}\|y_i\| + \sum_{i \in B}\|y_i\|\right)$$
$$\leq 2 + \frac{1}{r}(2k + r - k) \leq 3 + \frac{k}{r} \leq 3 + \frac{n_0}{r}.$$

Thus if $r \geq 2n_0$ then

$$\sum_{j=1}^{k} \|E_j \left(y_0 + \frac{1}{r}\sum_{i=1}^{r} y_i\right)\| \leq \frac{7}{2}. \tag{3}$$

Now by (2), (3), and the definition of the norm on T it follows that

$$\|y_0 + \frac{1}{r}\sum_{i=1}^{r} y_i\| \leq \frac{7}{4}, \text{ which contradicts (1)}.$$

Thus ℓ_1 does not embed isomorphically into T. □

Some basis theory needs to be recalled here, in the form of a "controlled" version of the Bessaga-Pelczynski Selection Principle:

Theorem I.4: Let X be a Banach space with basis $\{x_n\}_{n=1}^{\infty}$, and let $\{x_n^*\}_{n=1}^{\infty}$ be the associated sequence of coefficient functionals. If $\{y_n\}_{n=1}^{\infty} \subset X$ is a sequence such that

(i) $\varliminf_n \|y_n\| > 0$, and (ii) $\lim_n x_i^*(y_n) = 0, \forall i = 1, 2, \ldots,$

then, given $\epsilon > 0, \{y_n\}_{n=1}^{\infty}$ contains a basic subsequence which is $(1 + \epsilon)$-equivalent to a block basic sequence of $\{x_n\}_{n=1}^{\infty}$.

(We omit the proof, which can be found in [59].) We will use the above to produce the next few results.

Proposition I.5: T contains no subsymmetric basic sequences.

Proof. Suppose that T did contain a subsymmetric basic sequence $\{y_n\}_{n=1}^{\infty}$. ¿From the theory of Schauder bases, any subsymmetric basic sequence is either weakly null or equivalent to the canonical unit vector basis of ℓ_1. Proposition I.3 rules out the latter, so $y_n \to 0$ weakly. But then $\{y_n\}_{n=1}^{\infty}$ is bounded in the weak topology, hence norm bounded, and (by the Bessaga-Pelczynski Selection Principle) $\{y_n\}_{n=1}^{\infty}$ must have a subsequence $\{y_{n_i}\}_{i=1}^{\infty}$ which is equivalent to a normalized block basic sequence $\{x_i\}_{i=1}^{\infty}$ of $\{t_n\}_{n=1}^{\infty}$.

By passing to a subsequence, we may assume that

$$\mathrm{supp}\, x_i > i, \ \forall_i$$

By the assumption of subsymmetry, $\exists M > 0$ such that $\forall\, m \in \mathbb{N}, \forall$ scalars $a_1, a_2, \cdots, a_m$, and $\forall$ choices $n_1 < n_2 < \cdots < n_m$, we have:

$$\|\sum_{k=1}^{m} a_k y_k\| \geq M \|\sum_{k=1}^{m} a_k y_{n_k}\|.$$

But then,

$$
\begin{aligned}
\|\sum_{k=1}^{m} a_k y_k\| &\geq M \|\sum_{k=1}^{m} a_k y_{n_{m-1+k}}\| \\
&\geq M M' \|\sum_{k=1}^{m} a_k x_{m-1+k}\| \\
&\geq \tfrac{1}{2} M M' \sum_{k=1}^{m} |a_k|,
\end{aligned}
$$

where the constant M' comes from the equivalence of $\{x_i\}_{i=1}^{\infty}$ and $\{y_{n_i}\}_{i=1}^{\infty}$, and the final inequality comes from Proposition I.2(3). But this would imply that $\{y_k\}_{k=1}^{\infty}$ is equivalent to the unit vector basis of ℓ_1, an obvious contradiction. $\square$

Corollary I.6: T does not contain isomorphs of c_0 or ℓ_p, $(1 < p < \infty)$.

The standard bases of these spaces are subsymmetric. $\square$

Recall that the ℓ_p spaces $(1 < p < \infty)$ are uniformly convex, though c_0 is not. So, part of the above corollary can be had from the following:

Proposition I.7: T contains no infinite-dimensional uniformly convexifiable sub-spaces.

Proof. By the Bessaga-Peleczynski Selection Principle, every infinite-dimensional subspace of T has in turn (for given $\epsilon > 0$) a subspace which is $(1 + \epsilon)$-isomorphic to the span of a normalized block basic sequence in T, say $\{y_n\}_{n=1}^{\infty}$, where:

$$y_n = \sum_{i=p_n+1}^{p_{n+1}} a_i t_i, \text{ say.}$$

But by Proposition I.2 (3),

$$\sum_{i=n+1}^{2n} |\alpha_i| \geq \|\sum_{i=n+1}^{2n} \alpha_i y_i\| \geq \frac{1}{2} \sum_{i=n+1}^{2n} |\alpha_i|,$$

for all scalar choices $\{\alpha_i\}$, and all n.

Since ℓ_1^n cannot be embedded uniformly in a uniformly convexifiable space, no infinite-dimensional subspace of T is uniformly convexifiable. $\square$

We summarize all of this for the record in the following:

Theorem I.8: Tsirelson's space T is reflexive with a 1-unconditional basis and contains no isomorphic copies of c_0 or $\ell_p (1 \leq p < \infty)$, no subsymmetric basic sequences, and no uniformly convexifiable ("super-reflexive") subspaces of infinite dimension.

In particular, the unit vector basis $\{t_n\}_{n=1}^{\infty}$ is not subsymmetric, so it must have subsequences which are not equivalent to it. At this point in our study, we cannot explicitly identify any such subsequences, but in Chapter IV we give a lovely classification of them due to S. Bellenot [13].

The conclusions in the following proposition follow immediately from the definition of the norm in T (except for (4), which follows from (3)) and will be used (often implicitly) in much of what follows.

Proposition I.9:

1. If $x \in T$ has supp $x = E$, and if $m \geq 0$, then

$$\|x\|_{m+1} = \max \left\{ \|x\|_m, \frac{1}{2} \max \left[\sum_{j=1}^{k} \|(E \cap E_j)x\|_m \right] \right\},$$

   where the inner max is taken over all choices $k \leq E_1 < E_2 < \cdots < E_k, \ k = 1, 2, \cdots$.

2. If $x \in T, m \geq 0$, then

$$\|x\|_{m+1} = \max \left\{ \|x\|_m, \frac{1}{2} \sup \left[\sum_{j=1}^{k} \|E_j x\|_m \right] \right\},$$

   where the inner sup is taken over all choices

$$k \leq E_1 < E_2 < \cdots < E_k; k = 1, 2, \cdots, \text{for which}$$

$$(\max E_i) + 1 = \min E_{i+1}; \ i = 1, 2, \cdots, k - 1.$$

3. If $\{k_n\}_{n=1}^{\infty}$ and $\{j_n\}_{n=1}^{\infty}$ are two increasing sequences of natural numbers with $k_n \leq j_n, \forall n$, then $\|\sum_{n=1}^{\infty} a_n t_{k_n}\|_m \leq \|\sum_{n=1}^{\infty} a_n t_{j_n}\|_m$, for all choices of scalars $\{a_n\}_{n=1}^{\infty}$, and $\forall m \geq 0$ (In particular, $\{t_n\}_{n=1}^{\infty}$ is dominated by every subsequence of itself.)

4. If $\{k_n\}_{n=1}^{\infty}$ is an increasing sequence of natural numbers and if $\{t_{k_n}\}_{n=1}^{\infty}$ has a subsequence which is equivalent to $\{t_n\}_{n=1}^{\infty}$, then $\{t_{k_n}\}_{n=1}^{\infty}$ is equivalent to $\{t_n\}_{n=1}^{\infty}$.

The next proposition shows how the $(m + 1)$-norm of a vector in T is actually computed and will be useful in simplifying later arguments.

Proposition I.10: For any vector $x \in T$, and any $m \geq 0$, either

$$\|x\|_{m+1} = \sup \left\{ \frac{1}{2} \sum_{j=1}^{k} \|E_j x\|_m : k \leq E_1 < E_2 < \cdots < E_k, \ k = 1, 2, \cdots \right\},$$

or

$$\|x\|_{m+1} = \|x\|_0.$$

Proof. Suppose $x \in T, m \geq 0$, such that

$$\|x\|_{m+1} > \sup\left\{\frac{1}{2}\sum_{j=1}^{k}\|E_j x\|_m : k \leq E_1 < E_2 < \cdots < E_k, \ k = 1, 2, \cdots\right\}.$$

Then by the definition of $\|x\|_{m+1}$ we have $\|x\|_{m+1} = \|x\|_m$. It follows that

$$\|x\|_m > \sup\left\{\frac{1}{2}\sum_{j=1}^{k}\|E_j x\|_m : k \leq E_1 < E_2 < \cdots < E_k, \quad k = 1, 2, \cdots\right\}$$

$$\geq \sup\left\{\frac{1}{2}\sum_{j=1}^{k}\|E_j x\|_{m-1} : k \leq E_1 < E_2 < \cdots < E_k, \quad k = 1, 2, \cdots\right\},$$

and hence $\|x\|_m = \|x\|_{m-1}$. Continuing in this fashion, we conclude that $\|x\|_{m+1} = \|x\|_0$. $\square$

We now present one of the basic constructions in T from [17]. (The technique employed here is useful both in T and in other "Tsirelson-like" spaces.)

Lemma I.11: Fix a natural number m and an increasing sequence of natural numbers $k_1 < k_2 < \cdots$. Then for each vector $x = \sum_{n=1}^{\infty} a_n t_{k_n} \in \mathbb{R}^{(\mathbb{N})}$ there are sets $1 \leq F_1 < F_2 < F_3$ such that

$$\|\sum_{n=1}^{\infty} a_n t_{k_{2n}}\|_m \leq \sum_{\ell=1}^{3} \|F_\ell x\|_m. \tag{$*$}$$

Proof. We proceed by induction on m, with the case $m = 0$ being obvious. Now assume that $(*)$ holds for some $m \geq 0$ and for all vectors $x \in \mathbb{R}^{(\mathbb{N})}$. Choose $x = \sum_{n=1}^{\infty} a_n t_{k_n} \in \mathbb{R}^{(\mathbb{N})}$, and let $y = \sum_{n=1}^{\infty} a_n t_{k_{2n}}$. If $\|y\|_{m+1} = \|y\|_m$, we are done (by the inductive hypothesis). Otherwise there must exist a sequence $k_{2n} \leq E_1 < E_2 < \cdots < E_{k_{2n}}$ such that

$$\|y\|_{m+1} = \frac{1}{2}\sum_{j=1}^{k_{2n}}\|E_j y\|_m.$$

By applying the inductive hypothesis to each $E_j y$, we can produce sets $F_j^\ell, (j = 1, \cdots, k_{2n}; \ell = 1, 2, 3)$ such that $\|y\|_{m+1} \leq \frac{1}{2}\sum_{j=1}^{k_{2n}}\sum_{\ell=1}^{3}\|F_j^\ell y\|_m$, and such that $k_n \leq F_j^\ell, (j = 1, \cdots, k_{2n}; \ell = 1, 2, 3)$. We note that we lose no generality by assuming that the sets F_j^ℓ are mutually disjoint. (To see this, let f be the function $k_{2n} \to k_n$. Since y is supported on $\{t_{k_{2i}}\}_{i=1}^{\infty}$, we may assume

$$E_j \subset \{k_{2n}, k_{2(n+1)}, \cdots\}, \text{ for } j = 1, 2, \cdots, k_{2n}.$$

By the induction hypothesis, $\|E_j y\|_m \leq \sum_{\ell=1}^{e}\|F_j^\ell(E_j)x\|_m$, for some F_j^ℓ. Replacing F_j^ℓ by $F_j^\ell f(E_j)$ does the trick.)

It follows that:

$$k_n \leq F_1^1 < F_1^2 < F_1^3 < F_2^1 < \cdots < F_{k_{2n}}^1 < F_{k_{2n}}^2 < F_{k_{2n}}^3.$$

13

By reindexing these sets as $\{G_j : 1 \leq j \leq 3k_{2n}\}$ we may assume that

$$k_n \leq G_1 < G_2 < \cdots < G_{3k_{2n}}, \text{ and}$$

$$\|y\|_{m+1} \leq \frac{1}{2} \sum_{j=1}^{3k_{2n}} \|G_j x\|_m = \tag{1}$$

$$\frac{1}{2} \left(\sum_{j=1}^{k_n} + \sum_{j=k_n+1}^{k'_n} + \sum_{j=k'_n+1}^{3k_{2n}} \right) \|G_j x\|_m,$$

where $k'_n = k_{n+k_n}$.

$$\text{Let} \quad F_1 = \bigcup \ \{G_j : 1 \leq j \leq k_n\},$$
$$F_2 = \bigcup \ \{G_j : k_n + 1 \leq j \leq k'_n\},$$
$$F_3 = \bigcup \ \{G_j : k'_n + 1 \leq j \leq 3k_{2n}\}.$$

Then $k_n \leq F_1 < F_2 < F_3$. Since $k_n \leq G_1 < G_2 < \cdots < G_{k_n}$, by the definition of the norm in T, we have:

$$\frac{1}{2} \sum_{j=1}^{k_n} \|G_j x\|_m \leq \left(\bigcup_{j=1}^{k_n} G_j \right) x\|_{m+1} = \|F_1\|_{m+1}. \tag{2}$$

Since $k_n \leq G_1 < G_2 < \cdots < G_{k_n}$, and each G_j contains at least one k_s (for some s), it follows that $k'_n \leq G_{k_n+1} < G_{k_n+2} < \cdots < G_{k'_n}$. Again by the definition of the norm on T we have:

$$\frac{1}{2} \sum_{j=k_n+1}^{k'_n} \|G_j x\|_m \leq \| \left(\bigcup_{j=k_n+1}^{k'_n} G_j \right) x\|_{m+1} = \|F_2\|_{m+1}. \tag{3}$$

Exactly as above we may conclude that:

$$k_{n+k_n+k'_n} \leq G_{k'_n+1} < \cdots < G_{3k_{2n}}.$$

Since $n \leq k_n$, and $k_{2n} \leq k'_n$, we have that

$$3k_{2n} = k_{2n} + 2k_{2n} \ \leq k'_n + 2k_{n+k_n}$$
$$\leq k'_n + k_{n+k_n+k_{n+k_n}}$$

Therefore $3k_{2n} - k'_n \leq k_{n+k_n+k'_n}$ and by the definition of the norm on T again,

$$\frac{1}{2} \sum_{j=k'_n+1}^{3k_{2n}} \|G_j x\|_m \leq \| \left(\bigcup_{j=k'_n+1}^{3k_{2n}} G_j \right) x\|_{m+1} = \|F_3 x\|_{m+1}. \tag{4}$$

Now (1), (2), (3), and (4) imply:

$$\|y\|_{m+1} \leq \sum_{\ell=1}^{3} \|F_\ell x\|_{m+1}.$$

This completes the induction, and the proof. $\square$

By Proposition I.9 (3) and the above lemma, we get immediately:

Proposition I.12: For every increasing sequence of natural numbers $\{k_n\}_{n=1}^{\infty}$ and any choice of scalars $\{a_n\}_{n=1}^{\infty}$, we have:

$$\|\sum_{n=1}^{\infty} a_n t_{k_n}\| \leq \|\sum_{n=1}^{\infty} a_n t_{k_{2n}}\| \leq 3\|\sum_{n=1}^{\infty} a_n t_{k_n}\|.$$

In Chapter IV, we give a much stronger result than Proposition I.12, but we require this proposition in its proof. Proposition I.12 implies that every subsequence of $\{t_n\}_{n=1}^{\infty}$ spans a Banach space which is isomorphic to its Cartesian square. It also implies that any two subsequences $\{t_{k_i}\}_{i=1}^{\infty}$ and $\{t_{n_i}\}_{i=1}^{\infty}$ of $\{t_n\}_{n=1}^{\infty}$ have, in turn, further subsequences which are equivalent. (To see this, choose inductively natural numbers $i_i < i_2 < \cdots$ so that $d_{i_1} < n_{i_2} < k_{i_3} < n_{i_4} < \cdots$, and apply Proposition I.12 to the sequence $\left\{ t_{k_{i_1}}, t_{n_{i_2}}, t_{k_{i_3}}, t_{n_{i_4}}, \cdots \right\}$.) We also obtain:

Corollary I.13: For any increasing sequence of natural numbers $\{k_n\}_{n=1}^{\infty}$, $\{t_i\}_{i \neq k_{2n}}$ is 3-equivalent to $\{t_n\}_{n=1}^{\infty}$.

Proof: If $\{m_n\}_{n=1}^{\infty}$ is the increasing sequence of natural numbers which enumerates the set $\mathbb{N} \backslash \{k_{2n} : n = 1, 2, \cdots\}$, then it is easily checked that $m_n \leq 2n, \forall n$. Therefore by Propositions I.9(3) and I.12 for all sequences of scalars $\{a_n\}_{n=1}^{\infty}$, we have:

$$\left\| \sum_{n=1}^{\infty} a_n t_n \right\| \leq \left\| \sum_{n=1}^{\infty} a_n t_{m_n} \right\| \leq \left\| \sum_{n=1}^{\infty} a_n t_{2n} \right\| \leq 3 \left\| \sum_{n=1}^{\infty} a_n t_n \right\|.$$

$\square$

Definition I.14: Two bases $\{x_n\}_{n=1}^{\infty}$ and $\{y_n\}_{n=1}^{\infty}$ are _permutatively equivalent_ ("$\{x_n\} \overset{\approx}{\sigma} \{y_n\}$") if there is a permutation σ of $\mathbb{N}$ so that:

$$\{x_n\}_{n=1}^{\infty} \text{ is equivalent to } \{y_{\sigma(n)}\}_{n=1}^{\infty}.$$

(We write "$\{w_n\} \approx \{z_n\}$" to denote equivalence between bases $\{w_n\}_{n=1}^{\infty}$ and $\{z_n\}_{n=1}^{\infty}$.)

Remark I.15: Using Proposition I.12, we can show that for any $k_1 < k_2 \cdots$ that $\{t_n\}_{n=1}^{\infty}$ in T is permutatively equivalent to $\{(t_n, 0), (0, t_{k_n})\}_{n=1}^{\infty}$ in $T \oplus T$.

Proof: $\{m_n\}_{n=1}^{\infty}$ be the increasing enumeration for the set $\mathbb{N} \backslash \{k_n : n = 1, 2, \cdots\}$. (If this latter set is finite, Remark I.15 is trivial.) By Proposition I.12,

$$
\begin{aligned}
\{t_n\}_{n=1}^{\infty} \quad &\overset{\approx}{\sigma} \; \{(t_{m_n}, t_{k_n})\}_{n=1}^{\infty} \\
&\overset{\approx}{\sigma} \; \{(t_{m_n}, 0)(0, t_{k_n})\}_{n=1}^{\infty} \\
&\overset{\approx}{\sigma} \; \left\{ (t_{m_n}, 0)(0, t_{k_{2n}}), (0, t_{k_{2n+1}}) \right\}_{n=1}^{\infty} \\
&\overset{\approx}{\sigma} \; \{(t_{m_n}, 0), (t_{k_n}, 0), (0, t_{k_n})\}_{n=1}^{\infty} \\
&\overset{\approx}{\sigma} \; \{(t_n, 0), (0, t_{k_n})\}_{n=1}^{\infty}
\end{aligned}
$$

$\square$

Suppose $\exists$ a universal constant K such that $\{t_n\}_{n=1}^{\infty}$ is K-equivalent to $\{t_n\}_{n=k}^{\infty}, \forall k$. Then since $\{t_i\}_{i=k}^{2k-1}$ is 2-equivalent to the unit vector basis of ℓ_1^k, it would follow that $\{t_n\}_{n=1}^{k}$ is a $2K$-equivalent to the unit vector basis of $\ell_1^k, \forall k$. But this is impossible, since ℓ_1 does not embed into T.

Similarly, $\{t_n\}_{n=1}^{\infty}$ is not uniformly equivalent to $\{t_{kn}\}_{n=1}^{\infty}, \forall k$. That is: we cannot shift $\{t_n\}_{n=1}^{\infty}$ "too far" to the right, or "stretch it out" too much via a bounded operator. However, if we are

15

already k-units out in the sequence $\{t_n\}_{n=1}^\infty$ and we "stretch out" no more than k-units, we can do so via a bounded operator, i.e.:

Proposition I.16: $\forall k$, and for any scalars $\{a_n\}_{n=1}^\infty$,

$$\|\sum_{n=1}^\infty a_n t_{k-1+n}\| \le \|\sum_{n=1}^\infty a_n t_{kn}\|$$
$$\le 4\|\sum_{n=1}^\infty a_n t_{k-1+n}\|.$$

Proof: It suffices to prove this for $\mathbb{R}^{(\mathbb{N})}$. Fix $m \ge 0$ and let $k \ge 2$ be given. Pick j such that $2^{j-1} \le k < 2^j$. Let $x = \sum_{n=1}^\infty a_n t_{2^j n}$. Then $\|\sum_{n=1}^\infty a_n t_{kn}\| \le \|x\|$. Repeated application of Lemma I.11 shows that $\exists F_1 < F_2 < \cdots < F_{3^j}$ such that

$$\|x\|_m \le \sum_{\ell=1}^{3^j} \|F_\ell \sum_{n=1}^\infty a_n t_n\|_m$$
$$\le \sum_{\ell=1}^{3^j} \|F_\ell y\|_m, \text{ where } y = \sum_{n=1}^\infty a_n t_{k-1+n}.$$

Since $3^j < 2^{2j} \le 2^{j+1}k, \exists s \le j < k$ so that

$$2^s k + 1 \le 3^j < 2^{s+1} \cdot k.$$

We may assume that $k \le F_1$ (because of y's support). Hence, any sum of the form $\sum \{\|F_\ell y\|_m : 2^i \cdot k + 1 \le \ell \le 2^{i+1} \cdot k\}$ is an admissible sum for $2\|G_i y\|_{m+1}$, where $G_i = \bigcup \{F_\ell : 2^i k + 1 \le \ell \le 2^{i+1} \cdot k\}, 0 \le i \le s$, (and where $F_\ell = \phi$, for $\ell > 3^j$). (We abuse the usual notation for sums and unions here and throughout to help minimize secretarial suicide!) Then

$$\|x\|_m \le \left\{\sum_{\ell=1}^k + \sum_{\ell=k+1}^{2k} + \cdots + \sum_{\ell=2^s \cdot k+1}^{3^j}\right\} \|F_\ell y\|_m$$

$$\le 2\sum_{i=0}^s \|G_i y\|_{m+1} \le 4\|y\|_{m+2},$$

since $s < k \le G_0 < \cdots < G_s$. Letting m grow, we are done. $\square$

Repeated applications of the arguments of this propositions show that $\{t_n\}_{n=k+1}^\infty$ is 16-equivalent to $\{t_{2^k \cdot n}\}_{n=1}^\infty$, etc.

Notes and Remarks:

1/. Most of the results of this chapter have dual analogs in T^*. In particular, T^* is a reflexive Banach space with 1-unconditional basis $\{t_n^*\}_{n=1}^\infty$ (where t_n^* is the coefficient functional associated to t_n) which has no subsymmetric basic sequences and hence, no embedded copies of c_0 or any $\ell^p (1 \le p < \infty)$. Proposition I.2(3) becomes in T^*:

$$\sup_{i=1}^k |b_i| \le \|\sum_{i=1}^k b_i y_i^*\| \le 2\sup_{i=1}^k |b_i|.$$

Since we lack an analytic description of the norm in T^*, Propositions I.9 and I.10 do not have formulations in T^*, but by standard duality arguments we do get that:

$$\|\sum_n a_n t_{j_n}^*\| \ge \|\sum_n |; a_n t_{kn}^*\|,$$

for all sequences of scalars $\{a_n\}_{n=1}^{\infty}$ and all sequences of natural numbers $\{j_n\}_{n=1}^{\infty}$ and $\{k_n\}_{n=1}^{\infty}$, where $j_n \leq k_n, \forall n$.

Also any subsequence of $\{t_n^*\}_{n=1}^{\infty}$ which contains another subsequence equivalent to $\{t_n^*\}_{n=1}^{\infty}$ is itself equivalent to $\{t_n^*\}_{n=1}^{\infty}$.

Proposition I.12 dualizes to T^* with the corresponding inequality:

$$\frac{1}{3}\|\sum_n a_n t_{k_n}^*\| \leq \|\sum_n a_n t_{k_{2n}}^*\| \leq \|\sum_n a_n t_{k_n}^*\|.$$

Remark I.15 and its proof carry over to T^*. Corollary I.13 can be dualized to T^*, as can Proposition I.16.

2/. B. S. Tsirelson's original construction [56] actually yields the dual of the space denoted T in [26] and [35]. We can identify it with the dual of "our" T by a slight modification of [56]. Denote by Q_n the natural projection of T^* onto $[t_k^*]_{k=n}^{\infty}$ ("$[t_k^*]_{k=n}^{\infty}$", for the closed linear span of $\left\{t_n^*, t_{n+1}^*, \cdot\right\}$).

(In the original version, Q_n was onto $[t_k^*]_{k=n+1}^{\infty}$.)

Let B_{T^*} = the unit ball of T^*, and note that B_{T^*} has the following properties:

(a) $t_n^* \in B_{T^*}, \forall n$.

(b) $x \in B_{T^*}$ and $|y| \leq |x|$ imply that $y \in B_{T^*}$. (Here, $|\cdot|$ is the lattice structure induced by the unconditional basis.)

(c) If $\{y_i\}_{i=1}^{n}$ is a block basic sequence of B_{T^*}, then, $\frac{1}{2}Q_n\left(\sum_{i=1}^{n} y_i\right) \in B_{T^*}$.

In his paper, Tsirelson took A to be the smallest set in ℓ_{∞}, enjoying properties (a) through (c), and let K denote the closure of A with respect to the topology of pointwise convergence. Since it is easily seen that $K \subset c_0$, it follows that K is weakly compact. It is straightforward to check that the closed convex hull V of K defines a weakly compact convex set in c_0, which inherits properties (a) through (c). The minimality of A then forces

$$V = B_{T^*}.$$

3/. If $f : \mathbb{N} \to \mathbb{N}$ satisfies $f(n) < f(n+1), \forall n$, we can define a Tsirelson space T_f in the same manner as T with norm $\|| \cdot \||$, where the recursion becomes:

$$\||x\||_{m+1} = \max\left\{\||x\||_m, \frac{1}{2}\max\left[\sum_{j=1}^{f(k)} \||E_j x\||_m\right]\right\},$$

where the inner "max" is taken over all choices $k \leq E_1 < E_2 < \cdots < E_{f(k)}$, and $k = 1, 2, \cdots$. The following is now obvious from these definitions:

Proposition I.17: If $\{t_n'\}_{n=1}^{\infty}$ is the unit vector basis of T_f, then for all choices of scalars $\{a_n\}_{n=1}^{\infty}$ and all $m \geq 0$:

$$\|\sum_{n=1}^{\infty} a_n t_{f(n)}\||_m = \|\,\|\sum_{n=1}^{\infty} a_n t_n'\|\,\||_m.$$

Therefore $\{t_{f(n)}\}_{n=1}^{\infty}$ in T is isometrically equivalent to $\{t'_n\}_{n=1}^{\infty}$ in T_f. (In this notation, $T = T_I$, where I is the identity on $\mathbb{N}$.) Proposition I.12 implies (for $f(n) := 2n$) that $||| \cdot |||$ is equivalent to $\| \cdot \|$ on T. In particular, for a given f, $||| \cdot |||$ is an equivalent norm on T iff $\{t_{f(n)}\}_{n=1}^{\infty} \approx \{t_n\}_{n=1}^{\infty}$.

4/. J. Diestel relates to us that G. Fricke observed that T failed the Banach-Saks Property, but that T^* had it. Baernstein showed that Baernstein's space B was a reflexive Banach space without the Banach-Saks Property [8]. C. J. Seifert observed that B^* enjoyed the Banach-Saks Property [54]. T is the first reflexive Banach <u>lattice</u> known to fail the Banach-Saks Property.

5/. It's an easy exercise to show that the Tsirelson norm is the unique norm on $\mathbb{R}^{(\mathbb{N})}$ satisfying:

$$\|x\| = \max\left\{ \|x\|_0, \frac{1}{2}\sup\left\{ \sum_{i=1}^{k} \|E_i x\| : k \in \mathbb{N}, \{E_i\}_{i=1}^{k} \text{ admissible} \right\} \right\},$$

6/. Since T is reflexive, it has the Radon-Nikodym Property. It must fail super-R.N.P., since this would force T to be uniformly convexifiable.

7/. Some calculations suggest that for $x \in \mathbb{R}^{(\mathbb{N})}, \|x\| = \|x\|_{m(x)}$, where $m(x)$ is a natural number depending upon x. The function $m(x)$ has not been completely understood yet.

Chapter II: Block basic sequences in Tsirelson's space.

In this chapter we aim to carefully examine the rather special behavior of block basic sequences in Tsirelson's space. The techniques employed in many of the proofs here are often at least as important as the results themselves, both for the study of T and its many variations. Moreover, very little of the structure theory in later chapters is independent of the core of ideas presented here. Our main result: **every bounded block basic sequence of $\{t_n\}_{n=1}^{\infty}$ spans a complemented subspace of T and is equivalent to a subsequence of $\{t_n\}_{n=1}^{\infty}$** We build this result in stages, both for the purpose of clarity and because different parts of it generalize in different directions in later chapters. We begin with a result of P. Casazza, W. Johnson, and L. Tzafriri [17].

Lemma II.1: Let $y_n = \sum\limits_{i=p_n+1}^{p_{n+1}} a_i t_i, (n = 1, 2, \cdots)$ be a normalized block basic sequence of $\{t_n\}_{n=1}^{\infty}$. Then, for every sequence of scalars $\{b_n\}_{n=1}^{\infty}$,

$$\|\sum_n b_n t_{p_n+1}\| \leq \|\sum_n b_n y_n\|.$$

Proof: We will show that for all choices of scalars $\{b_n\}_{n=1}^{\infty}$:

$$\|\sum_n b_n t_{p_n+1}\|_m \leq \|\sum_n b_n y_n\|, \forall m. \tag{1}$$

The case $m = 0$ is trivial, since the y_n are normalized and the c_0-norm minorizes the Tsirelson norm. So assume that (1) holds for some fixed m and all $\{b_n\}_{n=1}^{\infty}$.

Let $x = \sum\limits_n b_n t_{p_n+1}$ and $y = \sum\limits_n b_n y_n$. Fix k, and let $k \leq E_1 < E_2 < \cdots < E_k$ be chosen, and consider the sum

$$\frac{1}{2}\sum_{j=1}^{k} \|E_j x\|_m. \tag{2}$$

Since supp $(x) \subset \{p_n + 1 : n \in \mathbb{N}\}$, we may as well assume that

$$E_j \subset \{p_1 + 1, p_2 + 1, \cdot\}, \forall j.$$

By applying the inductive hypothesis to each summand in (2), we obtain:

$$\frac{1}{2}\sum_{j=1}^{k} \|E_j x\|_m \leq \frac{1}{2}\sum_{j=1}^{k} \|\sum \{b_n y_n : p_n + 1 \in E_j\}\| \tag{3}$$

But since $p_n + 1 \in E_j$ implies $k \leq p_n + 1$, the sum on the right of (3) is admissible for y. Hence:

$$\frac{1}{2}\sum_{j=1}^{k} \|E_j x\|_m \leq \|y\|. \tag{4}$$

Now, by taking the supremum over all k and all $k \leq E_1 < E_2 < \cdots < E_k$ in (4), it follows from the inductive hypothesis and the definition of $\| \cdot \|_T$ that:

$$\|\sum_n b_n t_{p_n+1}\|_{m+1} \leq \|y\|.$$

This completes both the induction and the proof. $\square$

Corollary II.2: Every bounded block basic sequence of $\{t_n\}_{n=1}^{\infty}$ dominates $\{t_n\}_{n=1}^{\infty}$.

Proof: (The above Lemma, coupled with Proposition I.9 (3).) □

Note that bases with the property of Corollary II.2 are called <u>lower semi-homogeneous</u>, and were studied in [19]. Our next lemma (from [17]) shows that each bounded block basic sequence is dominated by a subsequence of $\{t_n\}_{n=1}^{\infty}$.

Lemma II.3: Let $y_n = \sum_{i=p_n+1}^{p_{n+1}} a_i t_i, (n = 1, 2, \cdots)$ be a normalized block basic sequence of $\{t_n\}_{n=1}^{\infty}$. Then for every sequence of scalars $\{b_n\}_{n=1}^{\infty}$ we have:

$$\|\sum_n b_n y_n\| \leq 6\|\sum_n b_n t_{p_{n+1}}\|.$$

Proof: Let $\||\cdot\||$ be the norm on T_f for $f(n) := 2n$, as described in (3) of the notes/remarks section of Chapter I, i.e., $\||\cdot\||$ sums $2n$ blocks beyond n instead of n blocks. By Propositions I.12 and I.17,

$$\|x\| \leq \||x\|| \leq 3\|x\|, \forall x \in \mathbb{R}^{(\mathbb{N})},$$

and hence, for any vector in T. To prove the lemma it suffices to show for any scalars $\{b_n\}_{n=1}^{\infty}$ and any $m \geq 0$ that:

$$x = \||\sum_n b_n y_n\||_m \leq 2\||\sum_n b_n t_{p_{n+1}}\||. \tag{1}$$

As usual, we proceed by induction on m, with the case $m = 0$ trivial. So assume that (1) holds for some fixed m and for all choices $\{b_n\}_{n=1}^{\infty}$. Let $\{b_n\}_{n=1}^{\infty}$ be arbitrary, and define

$$x = \sum_n b_n t_{p_{n+1}} \text{ and } y = \sum_n b_n y_n.$$

Choose any $k \leq E_1 < E_2 < \cdots < E_k$. By Proposition I.9(2), we may assume that

$$(\max E_j) + 1 = \min E_{j+1}, \forall j,$$

and (by enlarging E_k if necessary) that:

$$\max E_k = p_{j+1}, \text{ for some } j > k.$$

For each j and n, define:

$$F_n = \{p_n + 1, p_n + 2, \cdots, p_{n+1}\},$$
$$G_j = \bigcup\{F_n : F_n \subset E_j\},$$
$$H = \{n : F_n \cap E_j \neq \emptyset, \text{ for some } 1 \leq j \leq k, \text{ but } F_n \not\subset E_j\}.$$

Since the map $n \to \max\{j : 1 \leq j \leq k, F_n \cap E_j \neq \emptyset\}$ is increasing (and thus one-one when re-stricted to H), it follows that $\overline{\overline{H}} \leq k$. By the induction hypothesis and the definitions of $\|\cdot\|$

and $||| \cdot |||$, we have:

$$\frac{1}{2}\sum_{j=1}^{k} \|E_j y\|_m \leq \frac{1}{2}\sum_{j=1}^{k} \|G_j y\|_m + \frac{1}{2}\sum_{j=1}^{k}\sum_{\ell \in H} \|(E_j \bigcap F_\ell) y\|_m$$

$$\leq \frac{1}{2}\sum_{j=1}^{k} 2\||G_j x\|| + \frac{1}{2}\sum_{\ell \in H} 2\|F_\ell y\|_{m+1} \qquad (2)$$

$$\leq \sum_{j=1}^{k} \||G_j x\|| + \sum_{\ell \in H} \|F_\ell y\|.$$

For each $\ell \in H$,

$$\|F_\ell y\| = \|b_\ell y_\ell\| = |b_\ell|. \qquad (3)$$

It follows from (2) and (3) that:

$$\frac{1}{2}\sum_{j=1}^{k} \|E_j y\|_m \leq \sum_{j=1}^{k} \||G_j x\|| + \sum_{\ell \in H} \||b_\ell t_{p_{\ell+1}}\||. \qquad (4)$$

Since the right side of (4) contains at most $2k$ terms, it is an admissible sum for x in T_f. Thus:

$$\frac{1}{2}\sum_{j=1}^{k} \|E_j y\|_m \leq 2\||x\||. \qquad (5)$$

By taking the supremum of the left side of (5) over all choices $k \leq E_1 < E_2 < \cdots < E_k$ and all k, we complete the induction. $\square$

By Proposition I.12, if $y_n = \sum_{i=p_n+1}^{p_{n+1}} a_i t_i, (n = 1, 2, \cdots)$, is a normalized block basic sequence of $\{t_n\}_{n=1}^{\infty}$, then for all sequences of scalars $\{b_n\}_{n=1}^{\infty}$, we have:

$$\|\sum_n b_n t_{p_n+1}\| \leq \|\sum_n b_n t_{p_{n+1}}\| \leq 3\|\sum_n b_n t_{p_n+1}\|.$$

This fact, together with Lemmas II.1 and II.3, yields immediately:

Proposition II.4[17]: Let $y_n = \sum_{i=p_n+1}^{p_{n+1}} a_i t_i, (n = 1, 2, \cdots)$, be a normalized block basic sequence of $\{t_n\}_{n=1}^{\infty}$, Then for every choice of natural numbers $p_n < k_n \leq p_{n+1}, (n = 1, 2, \cdots)$, and every sequence of scalars $\{b_n\}_{n=1}^{\infty}$, we have:

$$\frac{1}{3}\|\sum_n b_n y_{2n-1}\| \leq \|\sum_n b_n y_{2n}\| \leq 18\|\sum_n b_n y_{2n-1}\|, \qquad (a)$$

and

$$\frac{1}{3}\|\sum_n b_n t_{k_n}\| \leq \|\sum_n b_n y_n\| \leq 18\|\sum_n b_n t_{k_n}\|. \qquad (b)$$

Corollary II.5: Let $\{E_n\}_{n=1}^{\infty}$ be a sequence of finite subsets of $\mathbb{N}$ such that $E_1 < E_2 < \cdots$, and let $k_n \in E_n, \forall n$. Then for each $x \in T$:

$$\frac{1}{3}\|\sum_n \|E_n x\| t_{k_n}\| \leq \|\sum_n E_n x\| \leq 18\|\sum_n \|E_n x\| t_{k_n}\|.$$

Proof: Apply Proposition II.4 to the block basic sequence $\{\|E_n x\|^{-1} \cdot E_n x\}_{n=1}^{\infty}, (E_n x \neq 0)$, with $b_n := \|E_n x\|$. $\square$

Corollary II.5 will be important in the next chapter for showing that T has the "blocking principle" of W. Johnson and M. Zippin [30,31]. A careful reading of the proofs of the results in [30,31] shows that it is the existence of this property which is crucial to them. (An explanation of these techniques can be found in the survey article on finite-dimensional decompositions [22].)

We can now prove one of the major results of this chapter:

Proposition II.6: Every block basic sequence of $\{t_n\}_{n=1}^\infty$ spans a complemented subspace of T.

Proof: Let $y_n = \sum_{i=p_n+1}^{p_{n+1}} a_i t_i, (n = 1, 2, \cdots)$, be a block basic sequence of $\{t_n\}_{n=1}^\infty$ For each n, we may choose $f_n \in T^*$ such that $f_n(y_n) = 1$ and $\|f_n\| = 1$. For each n, let $E_n = \{p_n + 1, p_n + 2, \cdots, p_{n+1}\}$. Now, define an operator $P : T \to [y_n]_{n=1}^\infty$ by:

$$Px = \sum_n f_n(E_n x) y_n, \forall x \in T.$$

For each n, m,

$$E_n y_m = \begin{cases} y_n, & \text{if } n = m, \\ 0, & \text{otherwise.} \end{cases}$$

Thus, $P = P^2$ is a linear projection of T onto $[y_n]_{n=1}^\infty$, and it suffices to prove that P is bounded.

But by Proposition II.4 and Corollary II.5, for all $x \in T$

$$\begin{aligned}
\|Px\| &\leq \|\sum_n |f_n(E_n x)| y_n\| \\
&\leq \|\sum_n \|E_n x\| y_n\| \\
&\leq 18\|\sum_n \|E_n x\| t_{p_n+1}\| \\
&\leq 54\|\sum_n E_n x\| \\
&\leq 54\|x\|.
\end{aligned}$$

$\square$

The version of the Bessaga-Pelczynski Selection Principle which we used in Chapter I yields that every subspace of T contains subspaces $(1 + \epsilon)$-close to the span of a block basic sequence against $\{t_n\}_{n=1}^\infty$. This fact (combined with Propositions II.4 and II.6 and standard arguments on perturbations of projections) yields:

Proposition II.7: Every subspace of T contains a complemented subspace isomorphic to $[t_{k_n}]_{n=1}^\infty$, for some $k_1 < k_2 < \cdots$. Hence, every basic sequence in T has a subsequence equivalent to a subsequence of $\{t_n\}_{n=1}^\infty$.

Notes and Remarks:

1/. Each result in this chapter has a dual version in T^*. We don't require any properties of T^*. to accomplish this, since it suffices to apply standard duality arguments to the corresponding results in T. As examples, the inequality of Lemma II.1 becomes in $T*$:

$$\|\sum_n b_n t_{p_n+1}^*\| \geq \|\sum_n b_n y_n^*\|,$$

while that of Lemma II.3 becomes:

$$\|\sum_n b_n y_n^*\| \geq \frac{1}{6}\|\sum_n b_n t_{p_{n+1}}^*\|.$$

Corollary II.2 (for T^*) claims that $\{t_n^*\}_{n=1}^\infty$ dominates all of its bounded block basic sequences. Bases with this property are called <u>upper semi-homogeneous</u>, and have been studied in [19]. We even have the T^*-analogues of Proposition II.4 and Corollary II.5. Please note that although the proof of Proposition II.6 was set in T, the result is actually fairly general, i.e., <u>any</u> Banach space with basis satisfying the conclusion of Proposition II.4 must also satisfy that of Corollary II.5, and hence must enjoy the conclusion of Proposition II.6. Similarly, Propositions II.4 and II.6 (and standard perturbation arguments) force Proposition II.7 to hold in any like Banach space, and hence in T^*.

2/. It's important to note here that the equivalence constants of a normalized block basic sequence in T with a subsequence of $\{t_n\}_{n=1}^\infty$ (as well as the norms of the projections of T onto these blocks) are block-independent.

3/. The results of this section form a fine demarcation in the theory of block basic sequences. Proposition II.4 shows that bounded block basic sequences of $\{t_n\}_{n=1}^\infty$ are equivalent to subsequences of $\{t_n\}_{n=1}^\infty$ and hence must dominate $\{t_n\}_{n=1}^\infty$, without necessarily being equivalent to $\{t_n\}_{n=1}^\infty$. This ought to be compared to a result of M. Zippin [60] which states that a basis $\{x_n\}_{n=1}^\infty$ which is equivalent to all of its block basic sequences must be equivalent to the canonical unit vector basis of c_0 or some $\ell_p(1 \leq p < \infty)$. J. Lindenstrauss and L. Tzafriri [37] have shown that if every block basic sequence of every permutation of an unconditional basis $\{x_n\}_{n=1}^\infty$ spans a complemented subspace of $[x_n]_{n=1}^\infty$, then $\{x_n\}_{n=1}^\infty$ is equivalent to the canonical unit vector basis of c_0 or some $\ell_p(1 \leq p < \infty)$. The first author and B. Lin [18] gave the first examples of unconditional bases for which every block basic sequence is complemented in the space, but for which the basis is not equivalent to the usual unit vector basis of c_0 or some ℓ_p, $(1 \leq p < \infty)$. Proposition II.6 shows that $\{t_n\}_{n=1}^\infty$ is another basis with this property.

Chapter III: Bounded linear operators on T and the "blocking" principle.

In this chapter we lay the groundwork for studying bounded linear operators on Tsirelson's space. We will also give a characterization of the isometries of Tsirelson's space, as well as a description of the action of the shift operator. We note here yet another striking property enjoyed by c_0, the ℓ_p spaces, and T: the "blocking" principle of W. Johnson and M. Zippin [30, 31]. We will state and prove some of these results in a setting slightly more general than Tsirelson's space, and refer the reader to [35] for basic information and nomenclature concerning finite- dimensional decompositions ("F.D.D.s").

Definition III.1: We say that a F.D.D. $\{C_n\}_{n=1}^{\infty}$ for a Banach space X is of type T (respectively: of type T^*) if $\exists$ natural numbers $1 \leq k_1 < k_2 \cdots$ such that for all choices of $x_n \in C_n, (n = 1, 2, \cdots)$, we have:

$$\sum_n x_n \text{ converges in } X$$

iff

$$\sum_n \|x_n\| t_{k_n} \text{ converges in } T(\text{respectively} : \sum_n \|x_n\| t_{k_n}^* \text{ converges in } T^*).$$

In standard Banach space notation, we would write such a space as $(\sum_n \oplus C_n)_{\{t_{k_n}\}_{n=1}^{\infty}}$, i.e., the direct sum of $\{C_n\}_{n=1}^{\infty}$ with respect to the basis $\{t_{k_n}\}_{n=1}^{\infty}$. Whenever it becomes important to specify the sequence $\{k_n\}_{n=1}^{\infty}$, we will say that " $\{C_n\}_{n=1}^{\infty}$ is of type T relative to $\{k_n\}_{n=1}^{\infty}$". The following is an immediate conclusion from definitions and the principle of uniform boundedness:

Proposition III.2: An F.D.D. $\{C_n\}_{n=1}^{\infty}$ for a Banach space X is of type T relative to $\{k_n\}_{n=1}^{\infty}$ iff $\exists$ constants $K, M > 0$ such that for any choices $x_n \in C_n, (n = 1, 2, \cdots)$, we have:

$$\frac{1}{M} \|\sum_n \|x_n\| t_{k_n}\|_T \leq \|\sum_n x_n\|_X$$
$$\leq K \|\sum_n \|x_n\| t_{k_n}\|_T.$$

It follows that (after a suitable renorming) we may assume:

$$\|\sum_n x_n\|_X = \|\sum_n \|x_n\| t_{k_n}\|_T. \tag{1}$$

Also (with this convention) note that Corollary II.5 becomes:;

Proposition III.3: Let $1 \leq E_1 < E_2 < \cdots$ be finite subsets of $\mathbb{N}$, and let $k_n \in E_n, (n = 1, 2, \cdots)$. If $C_n := [t_i : i \in E_n]$, then $\{C_n\}_{n=1}^{\infty}$ is an F.D.D. of type T (for its closed linear span in T) with respect to $\{k_n\}_{n=1}^{\infty}$.

(In particular, if $0 = p_1 < p_2 < \infty$, and $\forall n, E_n := \{p_n + 1, p_n + 2, \cdots, p_{n+1}\}$, then $\{C_n\}_{n=1}^{\infty}$ is an F.D.D. of type T for the space T, relative to $\{k_n\}_{n=1}^{\infty}$.)

Our next result shows that every blocking of an F.D.D. of type T is also an F.D.D. of type T.

Proposition III.4: Let $\{A_i\}_{i=1}^{\infty}$ be an F.D.D. of type T (respectively: of type T^* relative to $\{p(i)\}_{i=1}^{\infty}$. Let $B_n := [A_i : \ell_n \leq i < \ell_{n+1}]$ be a blocking of $\{A_i\}_{i=1}^{\infty}$ and let $\{k_n\}_{n=1}^{\infty}$ be chosen such that

$$p(\ell_n) \leq k_n \leq p(\ell_{n+1} - 1), \forall n.$$

Then $\{B_n\}_{n=1}^{\infty}$ is an F.D.D. of type T (respectively: of type T^* relative to $\{k_n\}_{n=1}^{\infty}$.

Proof: (We prove this only for T. All of the properties of T used in the proof also hold for T^*, so this suffices.)

We may assume without loss of generality that our F.D.D. satisfies equation (1) above. In particular, for a non-zero $y_n \in B_n$ of form

$$y_n = \sum \{x_i : \ell_n \leq i \leq \ell_{n+1} - 1\}, (x_i \in A_i),$$

we have

$$\|y_n\| = \| \sum \{\|x_i\| t_{p(i)} : \ell_n \leq i \leq \ell_{n+1} - 1\} \|.$$

By Proposition II.4 applied to each y_n, we obtain:

$$\frac{1}{3} \| \sum_n \quad \|y_n\| t_{k_n} \|$$
$$\leq \| \sum_n \|y_n\| \left(\|y_n\|^{-1} \cdot \sum \{\|x_i\| t_{p(i)} : \ell_n \leq i \leq \ell_{n+1} - 1\} \right) \|$$
$$\leq 18 \| \sum_n \|y_n\| t_{k_n} \|.$$

But (by (1)):

$$\| \sum_n \|y_n\| \quad \left(\|y_n\|^{-1} \cdot \sum \{\|x_i\| t_{p(i)} : \ell_n \leq i \leq \ell_{n+1} - 1\} \right) \|$$
$$= \| \sum_n \sum \{\|x_i\| t_{p(i)} : \ell_n \leq i \leq \ell_{n+1} - 1\} \|$$
$$= \| \sum_i \|x_i\| t_{p(i)} \| = \| \sum_i x_i \| = \| \sum_n y_n \|.$$

Thus by the definition, $\{B_n\}_{n=1}^{\infty}$ is an F.D.D. of type T relative to $\{k_n\}_{n=1}^{\infty}$. $\square$

We have now set the stage to give the basic technique for constructing bounded linear operators on spaces which have an F.D.D. of type T (respectively: T^* Since Proposition III.3 implies that T itself has an F.D.D. of type T, this result will also apply to T. The theorem is due to P. Casazza, W. Johnson, and L. Tzafriri [17].

Theorem III.5: Let $\{X_n\}_{n=1}^{\infty}$ be an F.D.D. of type T relative to $\{k_n\}_{n=1}^{\infty}$ for a Banach space X, and let $\{Y_n\}_{n=1}^{\infty}$ be an F.D.D. of type T relative to $\{j_n\}_{n=1}^{\infty}$ for a Banach space Y. Assume $\exists$ a natural number i such that:

$$k_n \leq j_{n+i} \text{ and } j_n \leq k_{n+i}, \forall n.$$

Moreover, suppose that $\{L_n : X_n \to Y_n\}_{n=1}^{\infty}$ are linear operators such that $\sup_n \|L_n\| < \infty$. Define (formally) $L : X \to Y$ by

$$L \sum_n x_n = \sum_n L_n x_n, (x_n \in X_n).$$

Then L is a bounded linear operator, with $\|L\|$ a function of i. If in addition, each L_n is one-to-one, and $\sup_n\|L_n^{-1}\| < \infty$, then L is an isomorphism of X into Y.

Proof: The assumptions on $\{k_n\}_{n=1}^\infty$ and $\{j_n\}_{n=1}^\infty$ and repeated applications of Proposition I.12 yield that $\{t_{j_n}\}_{n=1}^\infty$ is equivalent to $\{t_{k_n}\}_{n=1}^\infty$. Choose $K > 0$ such that

$$\|\sum_n b_n t_{j_n}\| \leq K\|\sum_n b_n t_{k_n}\|, \tag{2}$$

for all scalar sequences $\{b_n\}_{n=1}^\infty$. We may assume that both F.D.D.s satisfy equation (1), and so for $x_n \in X_n, (n = 1, 2, \cdot)$, we have:

$$\begin{aligned}
\|L(\sum_n x_n)\|_Y &= \|\sum_n L_n x_n\|_Y \\
&= \|\sum_n \|L_n x_n\| t_{j_n}\|_T \\
&\leq (\sup_n\|L_n\|)\|\sum_n \|x_n\| t_{j_n}\|_T \\
&\leq K(\sup_n\|L_n\|)\|\sum_n \|x_n\| t_{k_n}\|_T \\
&= K(\sup_n\|L_n\|)\|\sum_n x_n\|_X.
\end{aligned}$$

$\square$

It is easily seen that Theorem III.5 also holds in T^* since the under-pinnings of its proof hold both in T and T^*

Next we present the "blocking" principle for F.D.D.s of type T (or T^* (In Chapter VI, we will use this result to show that T^* is a minimal Banach space.) This theorem is also due to P. Casazza, W. Johnson, and L. Tzafriri [17].

Theorem III.6 (The "blocking" principle): Let $\{C_n\}_{n=1}^\infty$ be an F.D.D. for a subspace Y of a quotient space of a Banach space X, and assume that X has an F.D.D. $\{A_n\}_{n=1}^\infty$ which is of type T (respectively: of type T^*). Then $\{C_n\}_{n=1}^\infty$ has a blocking $\{D_i\}_{i=1}^\infty$ which is an F.D.D. of type T (respectively: of type T^*).

Proof: (By duality, it suffices to prove the theorem for T.) Assume Y is a subspace of a quotient space Z of X, and denote by $Q : X \to Z$ the quotient map. Let $\{C_n^*\}_{n=1}^\infty$ be the F.D.D. of Y^* determined by $\{C_n\}_{n=1}^\infty$. Let M be the F.D.D. constant of $\{C_n\}_{n=1}^\infty$, i.e., if $y_n \in C_n, \forall n$, and $k < j$, we have:

$$\|\sum_{n=1}^k y_n\| \leq M\|\sum_{n=1}^j y_n\|. \tag{3}$$

The dual Y^* of Y is isometric to a subspace of a quotient space W of X^*. Let $R : X^* \to W$ be the quotient map.

By [28], [30], and Proposition 1.g.4(b), [35] (and the remark following 1.g.4), applied successively to Q and R, we can construct blockings:

$$\begin{cases} B_i = [A_j : p(i) \leq j \leq p(i+1) - 1], & (i = 1, 2, \cdots), \\ \text{and} \\ B_i^* = [A_j^* : p(i) \leq j \leq p(i+1) - 1], & (i = 1, 2, \cdots), \end{cases}$$

of the given F.D.D. for X and the corresponding F.D.D. for X^*, and we can construct blockings:

$$\begin{cases} D_i = [C_j : q(i) \leq j \leq q(i+1) - 1], & (i = 1, 2, \cdots), \\ \text{and} \\ D_i^* = [C_j^* : q(i) \leq j \leq q(i+1) - 1], & (i = 1, 2, \cdots), \end{cases}$$

of the given F.D.D. for Y and the corresponding F.D.D. for Y^* so that:

$$\forall i, \forall y \in D_i, \exists x \in B_i \oplus B_{i+1} \text{ for which} \tag{a}$$
$$\|Qx - y\| \leq M^{-1} \cdot 2^{-(i+2)} \cdot \|y\|, \text{ and } \|x\| \leq 8\|y\|, \text{ and}$$

$$\forall i, \forall y^* \in D_i^*, \exists x^* \in B_i^* \oplus B_{i+1}^* \text{ for which} \tag{b}$$
$$\|Rx^* - y^*\| \leq M^{-1} \cdot 2^{-(i+2)} \cdot \|y^*\|, \text{ and } \|x^*\| \leq 8\|y^*\|.$$

Now let $y_i \in D_i, (i = 1, 2, \cdots)$, be a sequence of vectors which is eventually zero, and let $x_i \in B_i \oplus B_{i+1}$ be the corresponding sequence of vectors given by (a). Then:

$$\begin{aligned} \|\sum_i y_i\| &\leq \|\sum_i Qx_i\| + \sum_i \|y_i - Qx_i\| \\ &\leq \|\sum_i x_{2i}\| + \|\sum_i x_{2i-1}\| + M^{-1}\sum_i 2^{-(i+2)}\|y_i\|. \end{aligned}$$

By Proposition III.4, $\{B_n\}_{n=1}^\infty$ is an F.D.D. of type T relative to some sequence $\{k_n\}_{n=1}^\infty$. Using this and the fact that (3) implies

$$\|y_i\| \leq 2M\|\sum_k y_k\|, \ \forall i,$$

we obtain:

$$\|\sum_i y_i\| \leq K\|\sum_i \|x_{2i}\|t_{k_{2i}}\| + K\|\sum_i x_{2i-1}\|t_{k_{2i-1}}\| + \frac{1}{2}\|\sum_k y_k\|,$$

where K is the constant given by Proposition III.2 for the F.D.D. $\{B_n\}_{n=1}^\infty$. It follows by condition (a) and the 1-unconditionality of $\{t_n\}_{n=1}^\infty$ that:

$$\begin{aligned} \|\sum_i y_i\| &\leq 4K\|\sum_i \|x_i\|t_{k_i}\| \\ &\leq 32k\|\sum_i \|y_i\|t_{k_i}\|. \end{aligned}$$

By applying the same argument to the dual F.D.D., we have for all sequences $y_i^* \in D_i^*, (1 = 1, 2, \cdots)$, which are eventually zero:

$$\|\sum_i y_i^*\| \leq 4K\|\sum_i \|y_i^*\|t_{k_i}^*\|.$$

27

Therefore by standard duality arguments,

$$\left\|\sum_i y_i\right\| \geq (4KM)^{-1}\left\|\sum_i \|y_i t_{p(i)}\|\right\|.$$

It now follows that $\{D_i\}_{i=1}^\infty$ is an F.D.D. of type T. $\quad\square$

(The following result for T^* from [14] fails for F.D.D.s of type T, as we will see in Chapter VI.B.)

Theorem III.7: Every Banach space with an F.D.D. of type T^* embeds isomorphically into T^*.

Proof: Let $\{X_n\}_{n=1}^\infty$ be an F.D.D. of type T^* relative to $\{k_n\}_{n=1}^\infty$ for a Banach space X.

Since for each m, $\left[t_{k_n}^* : m \leq n \leq 2m-1\right]$ is 2-equivalent to the unit vector basis in ℓ_∞^m, and since ℓ_∞ is universal for all separable Banach spaces, we may choose $m(1) \in \mathbb{N}$ so that $W_0 := X_1$ is 2-isomorphic to a subspace of $\left[t_{k_n}^* : 1 \leq n \leq m(1)\right]$ and (as above) choose $m(2) > m(1)$ so that W_1 is 2-isomorphic to a subspace of $\left[t_{k_n}^* : m(1) < n \leq m(2)\right]$, where

$$W_1 := [X_n : 2 \leq n \leq m(1)].$$

Continuing in this fashion, we construct a blocking $\{W_i\}_{i=0}^\infty$ of $\{X_n\}_{n=1}^\infty$, and an increasing sequence $\{m(i)\}_{i=0}^\infty$, $(m(0) := 0)$, such that for $\forall i > 2$,

$$W_i = [X_n : m(i-1) < n \leq m(i)]$$

is 2-isomorphic to a subspace of $\left[t_{k_n}^* : m(i) < n \leq m(i+1)\right]$.

By Proposition III.4, $\{W_i\}_{i=0}^\infty$ is also an F.D.D. of type T^* for X, but relative to $\{k_{m(i)}\}_{i=1}^\infty$. Finally, $\forall i$, let L_i be an isomorphism from W_i onto a subspace of $\left[t_{k_n}^* : m(i) < n \leq m(i+1)\right]$ such that $\|L_i^{-1}\| = 1$ and $\|L_i\| \leq 2$. By Theorem III.5 and the discussion following it, the operator $L : X \to \left[t_{k_n}^*\right]_{n=1}^\infty$ defined for $x = \sum_{i=1}^\infty x_i \in X, (x_i \in X_i)$, by: $Lx = \sum_{i=0}^\infty L_i(x_i)$, is an isomorphism of X onto a subspace of T^*. $\quad\square$

The nature of certain sub-classes of operators from T to itself has been an on-going concern of many. The isometries of Tsirelson's space must of course include all coordinate-wise sign changes, as is easily seen by considering the construction of the norm. It turns out that there are very few other isometries. We have the following result of B. Beauzamy and P. G. Casazza, which appears here for the first time.

Theorem III.8: Any isometry U of T must have one of the following forms:

A $U t_n = \pm t_n$, for all $n \geq 1$, or

B $U t_1 = \pm t_2, U t_2 = \pm t_1, U t_n = \pm t_n$, for all $n \geq 3$.

Proof: The proof is long, though not complicated, relying primarily upon the definition of the norm. We will show the following:

28

i). any isometry leaves span $\{t_1, t_2\}$ invariant.

ii). on span $\{t_1, t_2\}$, the Tsirelson norm is the c_0-norm.

iii). for any isometry $U, Ut_3 = \pm t_3$.

iv). $Ut_k = \pm t_k, (k > 3)$.

v). the U's described in (B) of the Theorem are actually isometries.

Proof of i): We will say that " x starts beyond k", if $x_i = 0$, for $i \leq k$. Let U be any isometry, fix a_1, a_2 such that $|a_1| \leq 1, |a_2| \leq 1$. Write $U(a_1 t_1 + a_2 t_2) = a_1' t_1 + a_2' t_2 + x$, where x starts beyond 2. Consider $a_1 t_1 + a_2 t_2 + j_{j_1} + t_{j_2}$, where $2 < j_1 < j_2$. Clearly, we have: $\|a_1 t_1 + a_2 t_2 + t_{j_1} + t_{j_2}\| = 1$. Now $t_n \to 0$, weakly, thus so does Ut_n. Thus given $\epsilon > 0$, we can find j_i such that $a_1' t_1 + a_2' t_2 + x$ and Ut_{j_1} are "almost consecutive", i.e., letting $y_1 = Ut_{j_1}$, we can find x', a sub-block of x, and y_1', a sub-block of y_1 such that:

$$\|x - x'\| < e,$$
$$\|y_1 - y_1'\| < \epsilon, \text{ and}$$
$$x' \text{ and } y_1', \text{ are consecutive.}$$

With j_1 fixed this way, we can now find j_2 such that $y_2 = Ut_{j_2}$ is "almost after" y_1' : i.e., a sub-block y_2' of y_2 follows y_1' such that

$$\|y_2' - y_2\| < \epsilon.$$

Then we obtain:

$$1 + 3\epsilon \geq \|a_1' t_1 + a_2' t_2 + x' + y_1' + y_2'\|$$
$$\geq \tfrac{1}{2}(\|x'\| + \|y_1'\| + \|y_2'\|)$$
$$\geq \tfrac{1}{2}(\|x'\| + 2 - 2\epsilon).$$

Thus, $\|x\| \leq \|x - x'\| + \|x'\| < \epsilon + \|x'\| \leq \epsilon + 8\epsilon = 9\epsilon$, whence $x = 0$.

Proof of ii): This is clearly so, so we must have

$$\text{either } Ut_1 = \pm t_1, (\text{and then } Ut_2 = \pm t_2),$$
$$\text{or } Ut_1 = \pm t_2, (\text{and then } Ut_2 = \pm t_1).$$

Proof of iii): This we do in stages:

Step I: Write $Ut_3 = a_1 t_1 + a_2 t_2 + a_3 t_3 + x$, where x starts beyond 3. If $Ut_1 = \pm t_1$, we write $1 = \|U(t_1 \pm t_3)\| = \|(a_1 \pm 1)t_1 + a_2 t_2 + a_3 t_3 + x\| \geq |a_1 \pm 1|$, whence $a_1 = 0$. If $Ut_1 = \pm t_2$, then $Ut_2 = \pm t_1$, and by considering $t_2 \pm t_3$ we could again conclude $a_1 = 0$. A similar argument yields that $a_2 = 0$. Thus we obtain: $Ut_3 = a_3 t_3 + x$.

Step II: Next we show $Ut_3 = a_3 t_3 + a_4 t_4$, for some a_4. Writing $Ut_3 = a_3 t_3 + a_4 t_4 + x'$, where x' starts beyond 4, for $4 < j_1 < j_2 < j_3 < j_4$, we consider vectors of form $t_3 + t_{j_1} + t_{j_2} + t_{j_3} + t_{j_4}$, and note that they have norm =2. Letting $z_i = Ut_{j_i}, (i = 1,2,3,4)$, we obtain:

$$\|a_3 t_3 + a_4 t_4 + x' + z_1 + z_2 + z_3 + z_4\| = 2.$$

Choosing $\epsilon > 0, j_1 < j_2 < j_3 < j_4$ can be chosen so that z_1, z_2, z_3, z_4 are almost consecutive, and as before we obtain disjoint consecutive sub-blocks $z_1', \cdots, z_4'$ such that:

$$2 + 4\epsilon \geq \tfrac{1}{2}(\|x'\| + \|z_1'\| + \cdots + \|z_4'\|).$$
$$\geq \tfrac{1}{2}(\|x'\| + 4(1 - \epsilon)), \text{ so } \|x'\| \leq 12\epsilon,$$

whence $x' = 0$, just as before.

Step III: We prove that $Ut_3 = \pm t_3$ by showing $a_4 = 0$. Since $\|Ut_3\| = 1$, at least one of $|a_3|, |a_4|$ must be 1. But as before, for large j and $z_j = Ut_j$,

$$1 = \|t_3 + t_j\| = \|a_3 t_3 + a_4 t_4 + z_j\|$$
$$\geq \tfrac{1}{2}(|a_3| + |a_4| + 1 - \epsilon),$$

so $|a_3| + |a_4| = 1$, and one of $|a_3|, |a_4|$ must be 0. But a_3 cannot be 0, since if it were, $Ut_3 = \pm t_4$, and (fixing $\epsilon > 0$) for $j_1 < j_2 < j_3$ large enough:

$$\frac{3}{2} = \|t_3 + t_{j_1} + t_{j_2} + t_{j_3}\| = \|\pm t_4 + z_1 + z_2 + z_3\| \geq 2(1 - \epsilon),$$

a contradiction.

Proof of iv): Assume that we have proved that $Ut_k = \pm t_k$, for $3 \leq k < m$. We demonstrate $Ut_m = \pm t_m$ in three steps:

a. Ut_m does not begin before m:

write $Ut_m = a_1 t_1 + \cdots + a_m t_m + x$, where x starts beyond m. For $k < m$,

$$1 = \|Ut_m \pm Ut_k\| \geq |a_k \pm 1|, \text{ which implies } a_k = 0.$$

b. $Ut_m = a_m t_m + a_{m+1} t_{m+1}$, for appropriate scalars:

by (a), we can write $Ut_m = a_m t_m + a_{m+1} t_{m+1} + x'$, where x starts beyond $m + 1$. Choose $\epsilon > 0$. For $m < j_1 < \cdots < j_{m+1}$, (and as before, letting $z_\ell = Ut_{j_\ell}$) we have:

$$\frac{m+1}{2} = \|t_m + t_{j_1} + \cdots + t_{j_{m+1}}\|$$
$$= \|a_m t_m + a_{m+1} t_{m+1} + x' + z_1 + \cdots + z_{m+1}\|$$
$$\geq \tfrac{1}{2}(\|x'\| + \|z_1'\| + \cdots + \|z_{m+1}'\|)$$
$$\geq \tfrac{1}{2}(\|x'\| + (m + 1)(1 - \epsilon)), \text{ for}$$

$j_1 < \cdots < j_{m+1}$ large enough, and appropriate sub-blocks z_ℓ' of z_ℓ, and from this, $x' = 0$.

c. $Ut_m = +t_m$: since $\|t_m\| = 1$, at least one of $|a_m|, |a_{m+1}|$ must be 1. But

$$1 = \|t_m + t_j\|$$
$$= \|a_m t_m + a_{m+1} t_{m+1} + z_j\|,$$

from which we deduce (for large j):

$$1 + \epsilon \geq \frac{1}{2}(|a_m| + |a_{m+1}| + 1 - \epsilon),$$

and so one of a_m, a_{m+1} must be 0. But if $a_m = 0$, $Ut_m = \pm t_{m+1}$. So for $m < j_1 < j_2 \cdots < j_m$ sufficiently large,

$$
\begin{aligned}
\tfrac{m}{2} &= \|t_m + \sum_{i=1}^{m} t_{j_i}\| \\
&= \|t_{m+1} + \sum_{i=1}^{m} z_i\| \\
&\geq \|t_{m+1} + \sum_{i=1}^{m} z_i'\| - m\epsilon \\
&\geq \tfrac{m+1}{2} - m\epsilon, \text{ which is a contradiction for small } \epsilon.
\end{aligned}
$$

Hence, $a_m = \pm 1$, and $a_{m+1} = 0$, and finally we have:

$$Ut_k = \pm t_k, \quad (k > 3).$$

Proof of v): For this, it suffices to show:

$$\|a_1 t_1 + a_2 t_2 + x\| = \|a_2 t_1 + a_1 t_2 + x\|, \tag{$*$}$$

for all scalars a_1, a_2, and for all x which start beyond 2. Assume the left term in $(*)$ to be 1, the right ≤ 1. If $|a_1| = 1$, $(*)$ follows. Otherwise:

$$1 = \|a_1 t_1 + a_2 t_2 + x\| = \|a_2 t_2 + x\|.$$

If $|a_2| = 1$ or $\|x\|_{c_o} = 1$, equality also follows. Otherwise:

$$
\begin{aligned}
1 &= \|a_2 t_2 + x\| \\
&= \tfrac{1}{2}(\|E_1(a_2 t_2 + x)\| + \|E_2 x\|),
\end{aligned}
$$

for appropriate $2 \leq E_1 < E_2$.

But then $1 = \|E_1(a_2 t_2 + x)\| = \|E_2 x\|$, so $\|x\| = 1$, and $(*)$ is proved. □

Corollary III.9: All isometries of T are surjective.

In the sequence spaces c_o and $\ell_p (1 \leq p < \infty)$, we can define the shift operator S by:

$$S \sum_n a_n e_n = \sum_n a_n e_{n+1},$$

where $\{a_n\}_{n=1}^{\infty}$ are scalars and $\{e_n\}_{n=1}^{\infty}$ is the canonical unit vector basis. In these spaces S is an isometry, though by Corollary III.9 S is not an isometry on T. In general, $\|Sx\|_T \geq \|x\|_T, \forall x \in T$, since Sx has more admissible sums than x. (The example $\bar{x} = (0, 1, 1, 1, 1, 0, 0, \cdots) \in T$ shows that S can actually enlarge norm: $\|\bar{x}\| = \frac{3}{2}$, while $\|S\bar{x}\| = 2$.) A bit more can be said about the shift operator in T. The following is due to B. Beauzamy [10], who credits its proof to B. Maurey.

Proposition III.10: If $x \in \mathbb{R}^{(N)}$ with min supp $(x) =: n > 3$, then $\|S^{-1}x\| \geq \left(1 - \frac{3}{n}\right)\|x\|$.

Proof: Clearly, we may assume that $\|x\| = 1$. If $\|x\| = \|x\|_0$, then $\|S^{-1}x\| \geq \|S^{-1}x\|_o = \|x\|_0 = \|x\|$, and we're done.

Otherwise, $\exists$ an "admissible sum" for x, say $x = \sum_{i=1}^{\ell} x_i$, where $1 = \|x\| = \frac{1}{2}\sum_{i=1}^{\ell}\|x_i\|$, (i.e., $\{\text{supp } x_i\}_{i=1}^{\ell}$ is admissible).

There are two cases to consider:

I). $\{S^{-1}x_i\}_{i=1}^{\ell}$ is admissible, or

II). $\{S^{-1}x_i\}_{i=1}^{\ell}$ is not.

If we're in case I, we don't bother modifying $\{x_i\}_{i=1}^{\ell}$ at this stage. Case II occurs only if there is one summand too many. Now $l > n$, since x has no support before n. We "suppress" some $S^{-1}x_i$ of minimal norm, say $S^{-1}x_{i_0}$.

Surely, $\|S^{-1}x_{i_0}\| \leq \|x_{i_0}\|$, and since at least one of the $\|x_i\|$ must satisfy $\|x_i\| \leq \frac{2}{\ell}$, we get that:

$$\|S^{-1}x_{i_0}\| \leq \|x_{i_0}\| \leq \frac{2}{\ell} < \frac{2}{n}.$$

Restated:

$$\frac{1}{2}\|S^{-1}x_{i_0}\| \leq \frac{1}{n}.$$

Let $I_1 = \{1 \leq i \leq \ell : i \neq i_o\}$. If $i \in I_1$ and $\|x_i\| > \|x_i\|_0$, we write an admissible sum, say $\|x_i\| = \frac{1}{2}\sum_j \|x_{i,j}\|$, corresponding to i.

Letting $n_i = $ min supp (x_i), define length $(x_i) := (\text{ max supp } x_i) - n_i$.

If the family $\{S^{-1}x_{i,j}\}_j$ is inadmissible, then length $(x_i) \geq n_i$. Let $\{i_1 < i_2 < \cdots < i_p\}$ be the family of indices (if any) for which $\{S^{-1}x_{i,j}\}_j$ is inadmissible.

We have $n \leq n_{i_1}$, and since length $(x_{i_1}) \geq n_{i_1}$, we have

$$n_{i_2} \geq 2n_{i_1},$$
$$n_{i_3} \geq 2n_{i_2} \geq 2^2 n_{i_1}, \cdots,$$
$$n_{i_k} \geq 2^{k-1}n.$$

For each index $i_1, \cdots, i_p$ we "suppress" one of the elements of $\{S^{-1}x_{i_k,j}\}_j$ (of minimal norm), say $S^{-1}x_{i_k,j_x}$, and get that

$$\frac{1}{2}\|S^{-1}x_{i_k,j_k}\| \leq \frac{1}{2}\|x_{i_k,j_k}\| \leq \frac{1}{n_{i_k}} \leq 2^{1-k}n,$$

just as above.

The sum of the contributions to $\|x\|$ lost by "suppression" at this stage is (at most):

$$\frac{1}{2}\sum_{k=1}^{p}\left(\frac{1}{2}\|x_{i_k,j_k}\|\right) \leq \frac{1}{2}\left(\frac{1}{n} + \frac{1}{2n} + \frac{1}{4n} + \cdots\right) = \frac{1}{n}.$$

For the <u>next</u> stage, we let

$$I_2 = \{(i,j) : i \in I_1, \text{ with } j \text{ not already suppressed}\}.$$

If $\|x_{i,j}\| > \|x_{i,j}\|_0$ here, we write an admissible sum, say $\|x_{i,j}\| = \frac{1}{2}\sum_k \|x_{i,j,k}\|$.
We now list those $x_{i,j}$ for which $\{S^{-1}x_{i,j,k}\}_k$ is inadmissible as $z_1, z_2, \cdots, z_q$, say. Letting $\dot{n}_\ell = \min \text{supp}(z_\ell)$, as before, we have

$$\dot{n}_{\ell+1} \geq 2\dot{n}_\ell \geq 2^\ell n.$$

For each $z_\ell = x_{i,j}$ we "suppress" one of the $x_{i,j,k}$ (say z_{ℓ,k_ℓ}) for which

$$\frac{1}{2}\|x_{i,j,k}\| \leq \frac{1}{\dot{n}_\ell} \leq \frac{1}{2^{\ell-1}\cdot n}.$$

What we've "suppressed" at this stage has total norm at most

$$\frac{1}{4}\sum_\ell \|z_{\ell,k_\ell}\| \leq \frac{1}{4}\left(\frac{1}{n} + \frac{1}{2n} + \cdots\right) = \frac{1}{2n}.$$

By now, this process of "divide and conquer" should be clear, and the suppressed total at successive stages is no more than $\frac{1}{4n}, \frac{1}{8n}, \cdots$.

Since x has finite support, this process must eventually terminate (by bumping into the action of $\|\cdot\|_0$, and the total amount of norm eventually suppressed is at most $\frac{1}{n} + \frac{1}{n} + \frac{1}{2n} + \frac{1}{4n} + \cdots \leq \frac{3}{n}$.
Ignoring all of the suppressed vectors, we finally obtain an admissible sum for $S^{-1}x$ by shifting leftwards (by one index) the non-suppressed vectors which are admissible for x. It's clear that $\|S^{-1}x\| \geq 1 - \frac{3}{n}$. $\square$

Corollary III.11: For $n > 3$, the Banach-Mazur distance between $[t_i]_{i=n}^\infty$ and $[t_i]_{i=n+1}^\infty$ is $\leq \left(1 - \frac{3}{n}\right)^{-1}$.

Notes and remarks. 1/. It can be proved directly that Theorem III.7 fails for T, as follows. Let $X_n := \ell_\infty^n, n = 1, 2, \cdots)$, and let $X :=$ the Banach space with F.D.D. $\{X_n\}_{n=1}^\infty$ relative to $\{n\}_{n=1}^\infty$. For X to embed in T, we would need to have c_0 finitely representable in T. It's not difficult to see that this is impossible. What makes this result work in T^* is that all Banach spaces are finitely representable in c_0, that c_0 is finitely representable in T^* and the blocking principle. In Chapter IX, we prove the much stronger result that there is a function $f(x,y)$ with the property that a K-complemented subspace of T with an M-unconditional basis ($M < \infty$) must have a percentage of the basis which is $f(K,M)$-equivalent to the unit vector basis of ℓ_1^n, for some n. Since ℓ_∞^n is uniformly complemented in any Banach space in which it appears as a subspace, it follows that ℓ_∞^n does not embed uniformly in T.

2/. Tsirelson's space and its dual are the first examples besides c_0 and ℓ_p -F.D.D.s which have the "blocking" principle of W. Johnson and M. Zippin [30, 31].

3/. The most common application of Theorem III.5 is to the following case:

Let $\{E_n\}_{n=1}^{\infty}$ and $\{F_n\}_{n=1}^{\infty}$ be such that

$$
\begin{cases}
1 \le E_1 < E_2 < \cdots, \\
1 \le F_1 < F_2 < \cdots, \text{and} \\
E_{n-1} < F_n < E_{n+2}, \text{ for } n \ge 2.
\end{cases}
$$

Then Theorem III.5 applies to the F.D.D.s for T (respectively: for T^* of the form

$$
\begin{cases}
X_n = [t_i : i \in E_n], \\
Y_n = [t_i : i \in F_n].
\end{cases}
$$

$$
(\text{respectively} : \begin{cases}
X_n^* = [t_i^* : i \in E_n], \\
Y_n^* = [t_i^* : i \in F_n].)
\end{cases}
$$

Chapter IV. Subsequences of the unit vector basis of Tsirelson's space .

Since T contains no subsymmetric basic sequences, the canonical unit vector basis $\{t_n\}_{n=1}^{\infty}$ must have some subsequences which are not equivalent to it. Finding such subsequences is no easy task, and in fact we will demonstrate the existence of subsequences of $\{t_n\}_{n=1}^{\infty}$ with unbelievably large "gaps" in their support which are nonetheless equivalent to $\{t_n\}_{n=1}^{\infty}$. Their classification is due to S. Bellenot [13] and draws upon some results from [17].

A. Basic criterion for subsequences of $\{t_n\}_{n=1}^{\infty}$ to be equivalent to $\{t_n\}_{n=1}^{\infty}$.

The following is due to P. Casazza, W. Johnson, and L. Tzafriri:

Theorem IV.a.1 ([17]): Let $\{k_n\}_{n=1}^{\infty}$ be an increasing sequence in $\mathbb{N}$, and for each n, let $W_n := [t_i : k_n < i \leq k_{n+1}]$. Then the subsequence $\{t_{k_n}\}_{n=1}^{\infty}$ of $\{t_n\}_{n=1}^{\infty}$ is equivalent to $\{t_n\}_{n=1}^{\infty}$ iff $\sup_n \|I_n\| < \infty$, where I_n is the formal identity from W_n into ℓ_1 (i.e., $I_n(t_i) = e_i$, if $k_n < i \leq k_{n+1}$, and where $\{e_n\}_{n=1}^{\infty}$ is the canonical unit vector basis of ℓ_1).

Proof: First, we assume that $\{t_{k_n}\}_{n=1}^{\infty}$ is M-equivalent to $\{t_n\}_{n=1}^{\infty}$ for some $M \geq 1$. Fix n, and consider a vector of the form:

$$u_n = \sum \{a_i t_i : k_n < i \leq k_{n+1}\}.$$

Since $k_{k_{n+1}} - k_n \leq k_{k_n+1}$, it follows that $\frac{1}{2}\sum \{|a_i| : k_n < i \leq k_{n+1}\}$ is an admissible sum for $v_n := \sum \{a_i t_{k_i} : k_n < i \leq k_{n+1}\}$. Hence

$$\frac{1}{2}\sum \{|a_i| : k_n < i \leq k_{n+1}\} \leq \|v_n\| \leq M\|u_n\|.$$

It follows immediately that $\|I_n\| \leq 2M, \forall n$.

In order to prove the converse, we assume $\sup_n \|I_n\| =: C < \infty$.

Let $\begin{cases} j(0) = 0, \text{ and} \\ j(n) = k_{j(n-1)+1}, \text{ for } n \geq 1, \end{cases}$
and note that $j(0) < j(1) < j(2) < \cdots$.

$$\forall n, \text{ let } \begin{cases} E_n := \{i : j(n-1) < i \leq j(n)\}, \text{ and} \\ F_n := \{k_i : i \in E_n\}. \end{cases}$$

Then $\begin{cases} \max E_n = j(n), \min E_n = j(n-1)+1, \text{ and} \\ \max F_n = k_{j(n)}, \min F_n = k_{j(n-1)+1} = j(n). \end{cases}$

Therefore $E_{n-1} \leq j(n-1) < F_n < k_{j(n)+1} = j(n+1) < E_{n+2}, \forall n$, i.e., all of the conditions on $\{E_n\}_{n=1}^{\infty}$ and $\{F_n\}_{n=1}^{\infty}$ in remark (3) of the "Notes and remarks" of Chapter III are satisfied.

Let $U_n := [t_i : i \in E_n]$, and $V_n := [t_i : i \in F_n]$, and define an operator $L_n : U_n \to V_n$ by:

$$L_n\left(\sum \{a_i t_i : i \in E_n\}\right) = \sum \{a_i t_{k_i} : i \in E_n\},$$

for all choices of scalars $\{a_i : i \in E_n\}$, (and do this for each n).

35

Since $\bigcup_n E_n = \mathbb{N}$, the operator $L : T \to T$ defined formally by

$$Lx := \sum_n L_n(E_n x), \quad (x \in T),$$

has the property that $Lt_n = t_{k_n}, \forall n \in \mathbb{N}$. Since $\{t_{k_n}\}_{n=1}^{\infty}$ dominates $\{t_n\}_{n=1}^{\infty}$ these two basic sequences will be equivalent if L is bounded. By remark (3) of the "Notes and remarks" of the last chapter, L is bounded, if $\sup_n \|L_n\| < \infty$. This latter fact will now be demonstrated.

For every h, let

$$A_h := \left\{1 + k_{j(n-2)+h}, 2 + k_{j(n-2)+h}, \cdots, k_{j(n-2)+h+1}\right\}.$$

Then $E_n = \bigcup \{A_h : 1 \le h \le j(n-1) - j(n-2)\}$, for $n > 1$, and for every sequence of scalars $\{a_i : i \in E_n\}$, Lemma II.1 implies that:

$$\begin{aligned}
\left\| \sum \{a_i t_i : i \in E_n\} \right\| &= \left\| \sum \{\sum\{a_i t_i : i \subset A_h\} : 1 \le h \le j(n-1) - j(n-2)\} \right\| \\
&\ge \left\| \sum \{\|\sum\{a_i t_i : i \in A_h\}\| t_{1+k_{j(n-2)+h}} : 1 \le h \le j(n-1) - j(n-2)\} \right\|.
\end{aligned} \tag{1}$$

However by our assumption on the I_n's, we have:

$$\sum \{|a_i| : i \in A_h\} \le C \|\sum\{a_i t_i : i \in A_h\}\|, \text{ for each } h. \tag{2}$$

By (1),(2), and 1-unconditionality, we have:

$$\begin{aligned}
\left\| \sum \{a_i t_i : i \in E_n\} \right\| &\ge \\
C^{-1} &\left\| \sum \{\sum\{|a_i| : i \in A_h\} t_{1+k_{j(n-2)+h}} : 1 \le h \le j(n-1) - j(n-2)\} \right\|.
\end{aligned} \tag{3}$$

Now for $n \ge 1, j(n-1) - j(n-2) \le j(n-1) = k_{j(n-2)+1} \le 1 + k_{j(n-2)+h}$. Therefore the expression

$$\frac{1}{2} \sum \left\{\sum \{|a_i| : i \in A_h\} : 1 \le h \le j(n-1) - j(n-2)\right\}$$

is an admissible sum for the vector inside the norm on the right side of (3). Applying this fact to (3) yields:

$$\begin{aligned}
\|\sum \{a_i t_i : i \in E_n\|\| &\ge (2C)^{-1} \sum \{|a_i| : i \in E_n\} \\
&\ge (2C)^{-1} \|\sum\{a_i t_{k_i} : i \in E_n\}\|, \forall n.
\end{aligned}$$

Therefore $\|L_n\| \le 2C, \forall n > 1$, and, since L_1 is clearly a bounded operator, we can conclude that L is bounded. $\square$

The proof of Theorem IV.a.1 also works for subsequences of $\{t_n\}_{n=1}^{\infty}$:

Corollary IV.a.2: Let $\{t_{k_n}\}_{n=1}^{\infty}$ and $\{t_{j_n}\}_{n=1}^{\infty}$ be subsequences of $\{t_n\}_{n=1}^{\infty}$ and let

$$M_i = \{k_n : j_{i-1} < k_n \le j_i\}, \quad (i = 1, 2, \cdots), \quad (j_0 := 0),$$

and

$$N_i = \{j_n : k_{i-1} < j_n \le k_i\}, \quad (i = 1, 2, 3, \cdots), \quad (k_0 := 0).$$

Then $\{t_{k_n}\}_{n=1}^{\infty}$ is equivalent to $\{t_{j_n}\}_{n=1}^{\infty}$ iff $\sup_i \max\{\|I_i\|, \|J_i\|\} < \infty$, where I_i (respectively: J_i) is the formal identity from $[t_{j_n} : n \in N_i]$ (respectively: $[t_{k_n} : n \in M_i]$) into ℓ_1.

If $\{t_{j_n}\}_{n=1}^{\infty}$ is a subsequence of $\{t_{k_n}\}_{n=1}^{\infty}$ then $\bar{\bar{N}}_i = 1, \forall i$. Thus a subsequence $\{t_{j_n}\}_{n=1}^{\infty}$ of $\{t_{k_n}\}_{n=1}^{\infty}$ is equivalent to $\{t_{k_n}\}_{n=1}^{\infty}$ iff $\sup_i \|J_i\| < \infty$. In fact, if $k_n \leq j_n, \forall n$, then $\{t_{j_n}\}_{n=1}^{\infty}$ is equivalent to $\{t_{k_n}\}_{n=1}^{\infty}$ iff $\sup_i \|J_i\| < \infty$.

Notes and remarks.

1/. The results in this section all carry over to T^*, since the only theorems concerning T used in these proofs also hold in T^*. For example, Theorem IV.a.1 becomes:

Theorem IV.a.3: Let $\{k_n\}_{n=1}^{\infty}$ be an increasing sequence of natural numbers, and, for each n, let

$$W_n := [t_i^* : k_n < i \leq k_{n+1}\}.$$

Then the subsequence $\{t_{k_n}^*\}_{n=1}^{\infty}$ of $\{t_n^*\}_{n=1}^{\infty}$ is equivalent to $\{t_n^*\}_{n=1}^{\infty}$ iff $\sup_n \|I_n^{-1}\| < \infty$, where I_n is the formal identity map from W_n to c_0.

2/. It can also be shown that a subsequence $\{t_{k_n}\}_{n=1}^{\infty}$ of $\{t_n\}_{n=1}^{\infty}$ is equivalent to $\{t_n\}_{n=1}^{\infty}$ iff

$$\sup_n d\left([t_i : k_n < i \leq k_{n+1}], \ell_1^{m_n}\right) < \infty,$$

where $d(\cdot, \cdot)$ is the Banach-Mazur distance function, and $m_n = k_{n+1} - k_n$.

3/. At this point, we can list certain subsequences of $\{t_n\}_{n=1}^{\infty}$ which are not equivalent to $\{t - n\}_{n=1}^{\infty}$. Since $\{t_n\}_{n=1}^{\infty}$ has no subsequences which are equivalent to the unit vector basis of ℓ_1, for each M and $k, \exists n$ such that $\|I_{k,n}\| \geq M$, where $I_{k,n}$ is the formal identity map from $[j_i : k < i \leq n]$ to ℓ_1^{n-k}. Repeated applications of this result produce a sequence of natural numbers $0 = k_0 < k_1 < k_2 < \cdots$ such that $\|I_{k_{n-1}, k_n}\| \geq n, \forall n$. Theorem IV.a.1 now implies that $\{t_{k_n}\}_{n=1}^{\infty}$ is not equivalent to $\{t_n\}_{n=1}^{\infty}$. The problem with this argument is that it is <u>not</u> constructive, i.e., at this point we do not know for fixed k and M how large n must be to ensure that $\|I_{k,n}\| \geq M$. We will make this calculation in section C of this chapter.

4/. It's important to note in Theorem IV.a.1 that the equivalence constant between $\{t_n\}_{n=1}^{\infty}$ and $\{t_{k_n}\}_{n=1}^{\infty}$ is a function of $\sup_n \|I_n\|$. This allows us (later on) to choose a subsequence $\{t_{k_n}\}_{n=1}^{\infty}$ of $\{t_n\}_{n=1}^{\infty}$ so that $d \leq \|I_n\| \leq d + 1$, for a prescribed d, and $\forall n$. It follows that for any $m, \{t_{k_n} : n \geq m\}$ is equivalent to $\{t_n\}_{n=1}^{\infty}$ while the equivalence constant between them may be fixed to be as large as we want.

B. "Fast growing" subsequences of $\{t_n\}_{n=1}^{\infty}$ which are equivalent to $\{t_n\}_{n=1}^{\infty}$.

In this section we describe subsequences $\{t_{k_n}\}_n^{\infty}$ of $\{t_n\}_{n=1}^{\infty}$ with $k_{n+1} - k_n$ extremely large which are nevertheless equivalent to $\{t_n\}_{n=1}^{\infty}$. To facilitate this, we introduce a compact notation from logic.

The <u>fast growing hierarchy</u> [55] is a family of IN-valued functions on the natural numbers defined inductively for $n \geq 1$ via:

$$\begin{cases} g_0(n) = n + 1, \text{ and} \\ g_{i+1}(n) = g_i^n(n), \text{ for } i \geq 0, \end{cases} \tag{1}$$

where f^n denotes the n-fold iteration of f (with $f^0(k) := k$). It follows that

$$\begin{cases} g_1(n) = 2n, \\ g_2(n) = n\, 2^n, \end{cases}$$

and $g_3(n)$ is a "stacked tower" obtained by n-fold continuation of the following:

$$n\, 2^n, n\, 2^n\, 2^{(n\, 2^n)}, n \cdot 2^n \cdot 2^{(n\, 2^n)} \cdot 2^{n(n2^n) \cdot 2^{(n2^n)}}, \cdots.$$

We can define these functions for each countable ordinal α via:

a). If $\alpha = \beta + 1$, then the recursion (1) defines g_α in terms of g_β.

b). If α is a limit ordinal, let $\{\alpha(n)\}_{n=1}^\infty$ be the "natural" sequence of ordinals for which $\lim_n \alpha(n) = \alpha$. Now define $g_\alpha(n) = g_{\alpha(n)}(n)$.

In particular, $g_\omega(n) = g_n(n)$, where ω is the first uncountable ordinal.

As was already observed, $g_3(n)$ is rather large in comparison to n. Our first proposition [13] shows that for any $i \geq 2, \{t_{g_i(n)}\}_{n=1}^\infty$ is equivalent to $\{t_n\}_{n=1}^\infty$. To simplify the notation, we write

$$\begin{cases} \|x\|_T, \text{ for the Tsirelson norm of the vector } x = \sum_n a_n t_n, \text{ and} \\ \|x\|_1, \text{ for } \sum_n |a_n|. \end{cases}$$

Proposition IV.b.1: For all $i \geq 0, n \geq 2$, and any $x \in [t_j : n \leq j < g_{i+1}(n)]$, we have:

$$\|x\|_T \geq 2^{-(i+1)} \cdot \|x\|_1 \tag{2}$$

Proof. We use induction on i to demonstrate (2). If $i = 0, g_{i+1}(n) = 2n$, and the right side of (2) becomes $\frac{1}{2}\|x\|_1$, which is an admissible sum for $x \in [t_j : n \leq j < 2n]$. Thus (2) holds for $i = 0$.

Now assume (2) holds for some i, and let $x \in [t_j : n \leq j < g_{i+2}(n)]$. For each $1 \leq j \leq n$, let

$$E_j := \left\{ g_{i+1}^{j-1}(n), g_{i+1}^{j-1}(n) + 1, \cdots, g_{i+1}^j(n) - 1 \right\}.$$

Since $n \leq E_2 < \cdots E_n$, we have

$$\|x\|_T \geq \frac{1}{2} \sum_{j=1}^n \|E_j x\|_T.$$

Applying the inductive hypothesis to each summand $\|E_j\|_T$, we are done. □

Corollary IV.b.2: For each fixed $i \geq 0, \{t_{g_i(n)}\}_{n=1}^\infty$ is equivalent to $\{t_n\}_{n=1}^\infty$.

Proof. For $i \geq 0, n \geq 1$, we have $g_i(n) \geq n$, so by Proposition IV.b.1, if $x \in [t_j : g_i(n) \leq j < g_i(n+1)]$, then

$$\|x\|_1 \leq 2^{i+1} \cdot \|x\|_T \leq 2^{i+1} \|x\|_1.$$

38

So if I_n is the formal identity map from $[t_j : g_i(n) \leq j < g_i(n+1)]$ into ℓ_1, then $\|I_n\| \leq 2^{i+1}$. Theorem IV.a.1 can now be invoked. □

Notes and remarks.

1. For $i = 1$, Corollary IV.b.2 is the same as Proposition I.12.

2. By duality, Proposition IV.b.1 holds in T^*. In this setting, (2) becomes:

$$\|x\|_{T_*} \leq 2^{i+1}\|x\|_\infty, \text{ where } x \in \left[t_j^* : n \leq j < g_{i+1}(n)\right]. \tag{2'}$$

We also clearly have Corollary IV.b.2 for T^*.

3. In section C of this chapter, we will show that there are vectors $x \in T$ which "almost" produce equality in inequality (2) of Proposition IV.b.1. This will allow us to make a quantitative statement about the norm of the formal identity from $[t_j : g_i(n-1) < j \leq g_i(n)]$ into ℓ_1, and this statement will permit us to classify all subsequences of $\{t_n\}_{n=1}^\infty$ which are equivalent to $\{t_n\}_{n=1}^\infty$

4. The subsequences of $\{t_n\}_{n=1}^\infty$ which are made equivalent to $\{t_n\}_{n=1}^\infty$ by Corollary IV.b.2 have a fast rate of growth attached to their indices. Consider for example the following heirarchy:

$$\text{let } \begin{cases} \exp_0(n) := n, & (n \geq 1), \text{ and} \\ \exp_j(n) := 2^{\exp_{j-1}(n)}, & (j, n \geq 1). \end{cases}$$

Then $\exp_n(n) \leq g_3(n)$, as is easily seen, and hence $\{t_{\exp_n(n)}\}_{n=1}^\infty$ is equivalent to $\{t_n\}_{n=1}^\infty$

5. Although T has no subspace isomorphic to ℓ_1, Proposition IV.b.1 shows that the distance from ℓ_1^n to $[t_i]_{i=1}^n$ is surprisingly small. To quantify this statement, we introduce (recursively) another heirarchy of functions on the natural numbers:

$$\text{Set } \begin{cases} \log_0(n) := n, (n \geq 1) \\ \log_i(n) := \log(\log_{i-1}(n)), (n \text{ large enough so} \\ \qquad \text{that } \log_{i-1}(n) > 0), (i \geq 1). \end{cases}$$

With this notation, we can now show:

Proposition IV.b.3: For each fixed $m \geq 0$,

$$\lim_n \frac{d(\ell_1^n, [t_i]_{i=1}^n)}{\log_m(n)} = 0.$$

where $d(\cdot, \cdot)$ is the Banach-Mazur distance function.

Proof. For each n large enough, let $k := \log_{m+1}(n)$. Then, $g_{m+3}(k) \geq n$, and so by Proposition IV.b.1,

$$d(\ell_1^{n-k}, [t_i]_{i=k+1}^n) \leq 2^{m+3}.$$

Therefore, $d(\ell_1^n, [t_i]_{i=1}^n) \leq d(\ell_1^k, [t_i]_{i=1}^k) \cdot d(\ell_1^{n-k}, [t_i]_{i=k+1}^n) \leq k \cdot 2^{m+3}.$

It follows from the definition of k that:

$$\frac{d(\ell_1^n, [t_i]_{i=1}^n)}{\log_m(n)} \leq \frac{(\log_{m+1}(n)) \cdot 2^{m+3}}{\log_m(n)}.$$

Since it is clearly the case that

$$\lim_n \frac{\log_{m+1}(n)}{\log_m(n)} = 0,$$

we're done. □

In T^*, we have the following analog of Proposition IV.b.3:

Proposition IV.b.4: For each fixed $m \geq 1$,

$$\lim_n \frac{d(\ell_\infty^n, [t_i^*]_{i=1}^n)}{\log_m(n)} = 0.$$

C. Classification of the subsequences of $\{t_n\}_{n=1}^\infty$ which are equivalent to $\{t_n\}_{n=1}^\infty$.

Although S. Bellenot's classification [13] of the subsequences of $\{t_n\}_{n=1}^\infty$ which are equivalent to $\{t_n\}_{n=1}^\infty$ is a substantial contribution to our knowledge of Tsirelson's space, the lemmas upon which it is built (and their proofs) are equally important. In the discussion of these lemmas, we will continue following S.Bellenot's lead in avoiding the usual notation for summations. Earlier usages replacing $\sum_{i \in P} a_i t_i$ with $\sum \{a_i t_i : i \in P\}$ were typographically convenient. Here such notation becomes almost necessary.

Let us begin by stating the classification result:

Theorem IV.c.1: A subsequence $\{t_{k_n}\}_{n=1}^\infty$ of $\{t_n\}_{n=1}^\infty$ is equivalent to $\{t_n\}_{n=1}^\infty$ iff $\exists$ a natural number i such that $k_n \leq g_i(n)$, for all large n.

Theorem IV.c.1 shows, for example, that $\{t_{g_w(n)}\}_{n=1}^\infty$ is not equivalent to $\{t_n\}_{n=1}^\infty$. Note that the "if" part of the theorem is an immediate consequence of Proposition I.9 (3) and Corollary IV.b.2. The first lemma needed in the proof of the other half of Theorem IV.c.1 has some importance independent of the theorem, since by explicitly putting numbers into it, and tracing their action through the steps of the induction, we can glean the exact form of an element in $[t_i : n < i \leq g_\ell(n)]$ whose norm closely approximates $\|I_n\|$, where I_n is the formal identity from $[t_i : n < i \leq g_\ell(n)]$ into $\ell_1^k, (k := g_\ell(n) - n)$.

Lemma IV.c.2: $\forall n \geq 1, \ell \geq 0, \exists$ a sequence of scalars $\{a_k\}_{k=1}^\infty$ for which simultaneously:

(i). $\begin{cases} a_k = 0, \text{ for } k \leq n, \\ a_{n+1} = n^{-\ell}, \text{ and} \\ a_k > 0, \text{ for } k > n. \end{cases}$

(ii). $\sum \{a_k : m < k \leq g_\ell(m)\} = 1, (\text{for } m \geq n)$.

(iii). For $0 \leq j \leq \ell, f(m)$ is non-increasing, where

$$f(m) := \sum \{a_k : m < k \leq g_j(m)\}, \quad ;(\text{for } m > n).$$

$(iv).$ $\sum \{a_k : m(1) < k \leq m(2)\} = \sum \{a_k : g_\ell^j(m(1)) < k \leq g_\ell^j(m(2))\}$, (for $j \geq 0, n \leq m(1) < m(2)$).

$(v).$ $\dfrac{\sum\{a_k : g_{\ell+1}(p) < k \leq g_{\ell+1}(p+1)\}}{\sum\{a_k : g_{\ell+1}(p+1) < k \leq g_{\ell+1}(p+2)\}} \leq \dfrac{a_{p+1}}{a_{p+2}}$, $(p \geq n)$.

$(vi).$ $\sum \{a_k : n < k \leq g_j(n)\} \leq n^{j-\ell}$, $(0 \leq j \leq \ell)$.

Proof: We proceed by induction on ℓ. For $\ell = 0, a_k := 1$ solves the Lemma (for $k > n$). (In fact, by (ii), this is the only sequence which works in this case.) The inequality in (v) has $\frac{2}{2}$ for its left term, since $g_1(n) = 2n$. The other conditions are clearly satisfied.

Now assume that $\{a_k\}$ is the sequence for some fixed ℓ. We will construct a non-increasing sequence $\{b_k\}$ so that $\{a_k \cdot b_k\}$ works for $\ell + 1$.

$$\text{Let } b_k = \begin{cases} 1, & \text{for } k \leq n, \\ \frac{1}{n}, & \text{for } n < k \leq g_{\ell+1}(n). \end{cases}$$

The terms b_k for k beyond $g_{\ell+1}(n)$, are defined a "block at a time" by induction.

For $g_{\ell+1}(n) < k \leq g_{\ell+1}(n+1)$, choose b_k to be that constant K such that:

$$a_{n+1} \cdot b_{n+1} = K \cdot \sum \{a_k : g_{\ell+1}(n) < k \leq g_{\ell+1}(n+1)\}$$

In general, for $g_{\ell+1}(p) < q \leq g_{\ell+1}(p+1)$, choose b_q so that:

$$a_{p+1} \cdot b_{p+1} = b_q \cdot \sum \{a_k : g_{\ell+1}(p) < k \leq g_{\ell+1}(p+1)\}.$$

Now we proceed to check that $\{a_k \cdot b_k\}$ satisfies (i) through (vi) for $\ell + 1$.

Conditions (i) and (ii) are satisfied by the definition of the terms b_k.

Next we demonstrate that $\{b_k\}_{k=1}^{\infty}$ is a non-increasing sequence. The first thing to show is that:

$$b_{g_{\ell+1}(n)} \geq b_{g_{\ell+1}(n+1)}$$

Since $b_{g_{\ell+1}(n)} = \frac{1}{n}$, and $a_{n+1} \cdot b_{n+1} = n^{-\ell-1}$, if we can show that $\sum \{a_k : g_{\ell+1}(n) < k \leq g_{\ell+1}(n+1)\} \geq 1$, then it will follow that:

$$b_{g_{\ell+1}(n)} = n^{-1} \geq n^{-\ell-1} \geq b_{g_{\ell+1}(n+1)}.$$

Since $n + 1 > n$, and all the g are increasing functions,

$$\begin{cases} g_{\ell+1}(n) = g_\ell^n(n), \text{ and} \\ g_{\ell+1}(n+1) = g_\ell^{n+1}(n+1) \geq g_\ell^{n+1}(n). \end{cases}$$

Thus by (ii),

$$\sum \{a_k : g_{\ell+1}(n) < k \leq g_{\ell+1}(n+1)\| \geq \sum \{a_k : g_\ell^n(n) < k \leq g_\ell^{(n+1)}(n)\} = 1.$$

Next assume $g_{\ell+1}(p) < q \leq g_{\ell+1}(p+1) < r \leq g_{\ell+2}(p+2)$ are chosen.

Then $\begin{cases} a_{p+1} \cdot b_{p+1} = b_q \cdot s_1, \text{ and} \\ a_{p+2} \cdot b_{p+2} = b_r \cdot s_2, \text{ where} \end{cases}$

$$\begin{cases} s_1 := \sum \{a_k : g_{\ell+1}(p) < k \le g_{\ell+1}(p+1)\}, \\ s_2 := \sum \{a_k : g_{\ell+1}(p+1) < k \le g_{\ell+1}(p+2)\}, \end{cases}$$

So $b_q \ge b_r$ iff $\frac{a_{p+1} \cdot b_{p+1}}{s_1} \ge \frac{a_{p+2} \cdot b_{p+2}}{s_2}$, or, equivalently,

$$1 \ge \frac{a_{p+2} \cdot b_{p+2} \cdot s_1}{a_{p+1} \cdot b_{p+1} \cdot s_2}.$$

But by condition (v) for $\{a_k : k > n\}$,

$$\frac{a_{p+2} \cdot b_{p+2}}{a_{p+1} \cdot b_{p+1}} \cdot \frac{s_1}{s_2} \le \frac{a_{p+2} \cdot b_{p+2}}{a_{p+1} \cdot b_{p+1}} \cdot \frac{a_{p+1}}{a_{p+2}} = \frac{b_{p+2}}{b_{p+1}}.$$

Since $\{b_k\}$ is known to be non-increasing for $k \le q$, it follows that: $\frac{b_{p+2}}{b_{p+1}} \le 1$. Therefore $\{b_k\}_{k=1}^{\infty}$ is a non-increasing sequence.

Next we show that (iii) holds for $\ell + 1$. Note that condition (iii) for ℓ is equivalent to:

$$0 \le f(m) - f(m+1) = a_{m+1} - \sum \{a_k : g_j(m) < k \le g_j(m+1)\}.$$

Also since $b_{m+1} \ge b_k$, for all $g_j(m) < k \le g_j(m+1)$, this condition still holds for $\ell + 1$ and for all $j \le \ell$. Since (ii) gives it immediately for $j := \ell + 1$, (iii) holds for $\ell + 1$.

Now we demonstrate (iv) for $\ell + 1$. Note first that by (ii):

$$\sum \{a_k b_k : m < k \le g_{\ell}(m)\} = 1 = \sum \{a_k b_k : m < k \le g_{\ell}(m+1)\}.$$

Subtracting off the common terms yields:

$$a_{m+1} \cdot b_{m+1} = \sum \{a_k b_k : g_{\ell}(m) < k \le g_{\ell}(m+1)\}.$$

Thus:

$$\begin{cases} a_{m(1)+1} \cdot b_{m(1)+1} = \sum \{a_k b_k : g_{\ell}(m(1)) < k \le g_{\ell}(m(1)+1)\}, \\ a_{m(1)+2} \cdot b_{m(1)+2} = \sum \{a_k b_k : g_{\ell}(m(1)+1) < k \le g_{\ell}(m(1)+2)\}, \cdots, \\ a_{m(2)} \cdot b_{m(2)} = \sum \{a_k b_k : g_{\ell}(m(2)-1) < k \le g_{\ell}(m(2))\}, \end{cases}$$

Summing these equalities yields (iv) when $j = 1$. Straightforward induction on j now yields (iv).

Now we show (v) for $\ell + 1$. By (iv),

$$\begin{aligned} &\sum \{a_k b_k : g_{(\ell+1)}(p) < k \le g_{\ell+1}(p+1)\} = \\ &\sum \{a_k b_k : g_{\ell}^p(p) < k \le g_{\ell}^p(g_{\ell}(p+1))\} = \\ &\sum \{a_k b_k : p < k \le g_{\ell}(p+1)\}. \end{aligned} \tag{1}$$

Similarly,

$$\begin{aligned} &\sum \{a_k b_k : g_{\ell+1}(p+1) < k \le g_{\ell+1}(p+2)\} = \\ &\sum \{a_k b_k : p < k \le g_{\ell}(p+1)\}. \end{aligned} \tag{2}$$

If $x := \sum\{a_k b_k : p+1 < k \le g_\ell(p+1)\}$, then by virtue of (1),

$$\sum\{a_k b_k : g_{\ell+1}(p) < k \le g_{\ell+1}(p+1)\} = a_{p+1} \cdot b_{p+1} + x.$$

Also by (2) and (iv),

$$\sum\{a_k b_k : g_{\ell+1}(p+1) < k \le g_{\ell+1}(p+2)\} =$$
$$x + \sum\{a_k b_k : g_\ell(p+1) < k \le g_\ell(p+2)\} = x + a_{p+2} \cdot b_{p+2}.$$

Since $a_{p+1} b_{p+2} > a_{p+2} b_{p+2}$, it follows that $f(x) := \frac{a_{p+1} b_{p+1} + x}{a_{p+2} b_{p+2} + x}$ is non-increasing for $x \ge 0$. This completes the proof of (v) for $\ell + 1$.

Finally we show that (vi) holds for $\ell + 1$. By the definition of the terms b_k,

$$\sum\{a_k b_k : n < k \le g_{\ell+1}(n)\} = \frac{1}{n}\sum_{j=1}^n \sum\{a_k b_k : g_\ell^{j-1}(n) < k \le g_\ell^j(n)\} = 1.$$

So if $j \le \ell$,

$$\sum\{a_k b_k : n < k \le g_j(n)\} = \frac{1}{n}\sum\{a_k : n < k \le g_j(n)\} \le \frac{1}{n} \cdot n^{j-\ell} = n^{j-(\ell+1)}.$$

Thus, (vi) is obtained for $\ell + 1$, and finally the lemma is demonstrated. $\square$

Using Lemma IV.c.2, we can get an estimate on the best equivalence constant between $\{t_k : n < k \le g_{\ell+1}(n)\}$ and the unit vector basis of ℓ_1^m, where $m := g_{\ell+1}(n) - n$. The proof of this lemma is important, because it yields the first systematic way of tackling what had long seemed an insurmountable difficulty, namely, given $x \in T$, reducing $\|x\|_m$ to a sum of $(m-1)$-st norms, and then iterating this process down to 0-th norms. The difficulties of this process are ironed out in the proof of the following:

Proposition IV.c.3: For each $\ell \ge 0$ and each $n \ge 2$, $\exists$ a non-zero $x = \sum\{a_k t_k : n < k \le g_{\ell+1}(n)\}$ such that

$$\|x\| \le \left(\sum_{j=1}^{\ell+1} 2^{-j} n^{j-\ell-1}\right) \sum\{|a_k| : n < k \le g_{\ell+1}(n)\}.$$

Proof. Fix $\ell \ge 0, n \ge 2$, and let $x = \sum\{a_k t_k : n < k \le g_\ell^n(n)\}$, where $\{a_k\}_{k=1}^\infty$ is the sequence given in Lemma IV.c.2 for ℓ. Since $g_{\ell+1}(n) = g_\ell^n(n)$, we need only show that this x satisfies the inequality of the proposition.

To simplify the proof, we introduce some terminology. A vector $y \in T$ is a <u>parent</u> if $\|y\|_T > \|y\|_0$. Otherwise y is a <u>leaf</u>. If y is a parent, its <u>children</u> are vectors $\{E_j y\}_{j=1}^k$ where

$$\begin{cases} \{E_j\}_{j=1}^k \text{ is admissible, and} \\ \|y\|_T = \frac{1}{2}\sum_{j=1}^k \|E_j y\|_T, \end{cases}$$

where we choose one such $\{E_j\}_{j=1}^k$ for each parent. Thus, with x as a <u>root</u>, we obtain a <u>tree</u> of vectors built upon the root x. The <u>level</u> of a vector in this tree is the number of parents of parents of $\cdots$ necessary to relate that vector to the root, i.e., the level 1 vectors in the x-tree are the children of x, the level 2 vectors are the "grandchildren" of x, etc.

Inductively, it's easy to show that if x is a parent, that

$$\|x\|_T = \sum_{j=1}^{k} 2^{-j} \left(\sum \{\|y\|_0 : y \text{ is a level } -j \text{ leaf}\} \right) \tag{3}$$
$$+ 2^{-(k+1)} \sum \{\|y\|_T : y \text{ is a level } -(k+1) \text{ vector}\}.$$

(Note that if x is not a parent, then:

$$\begin{cases} \|x\|_T = \|x\|_0 = a_{n+1} = n^{-\ell}, \text{ and} \\ \|x\|_{\ell_1} = n, \end{cases}$$

so if $n \geq 2$, the inequality of Proposition IV.c.3 holds.)

Now we assume that x is a parent and estimate the two sums in (3), assuming that $k \leq \ell$. Since $\|y\|_T \leq \|y\|_{\ell_1}$, and vectors on the same level are disjointly supported, it follows that:

$$\sum \{\|y\|_T : y \text{ is a level } -(k+1) \text{ vector}\} \leq \|x\|_{\ell_1} = n.$$

Next we estimate $\sum \{\|y\|_0 : y \in L_j\}$, where $L_j := \{y : y \text{ is a level } -j \text{ leaf}\}$, by tracing our way down the tree to the root. For $1 \leq i \leq j - 1$, let P_j^i be the set of level $-(j - i)$ vectors which have some $y \in L_j$ as a descendent, i.e.,

$$\begin{cases} P_j^1 \text{ the set of parents of elements of } L_j, \text{ and} \\ P_j^{i+1} \text{ is the set of parents of elements of } P_j^i. \end{cases}$$

Thus P_j^i is either void, or $\{x\}$.

We will prove, by induction on i, that for each $z \in P_j^i, \exists$ a natural number $m(z)$ satisfying the following three conditions:

(a). $\left\{ \{k : m(z) < k \leq g_i(m(z))\} : z \in P_j^i \right\}$ is pairwise disjoint,

(b). $\sum \{\|y\|_0 : y \in L_j\} \leq \sum \left\{ \sum \{a_k : m(z) < k \leq g_i(m(z))\} : z \in P_j^i \right\}$,

(c). $m(z) + 1 \geq \min \operatorname{supp}(z)$.

For the induction, the case $i = 0$ is obvious, since

$$\|y\|_0 = \sum \{a_k : m(y) < k \leq g_0(m(y))\} = a_{m(y)+1},$$

where $m(y) + 1 = \min \operatorname{supp}(y)$.

So assume that the claim is true for i, and select $w \in P_j^{i+1}$.

Let

$$\begin{cases} C(w) := \left\{ z \in P_j^i : z \text{ is a child of } w \right\}, \text{ and} \\ \\ m(w) := \min \{m(z); z \in C(w)\}, \end{cases}$$

and let $C(w)$ be listed as $\{z_1, z_2, \cdots, z_q\}$, where the terms z are ordered by $m(z_1) < m(z_2) < \cdots < m(z_q)$. Now by part (a) of the inductive hypothesis $g_i^r(m(w)) \leq m(z_{r+1})$. So by Lemma

44

IV.c.2 (which applies, since $j \le k \le \ell$, by assumption),

$$\sum \left\{ a_k : g_i^r(m(w)) < k \le g_i^{r+1}(m(w)) \right\} \ge$$
$$\sum \left\{ a_k : m(z_{r+1}) < k \le g_i(m(z_{r+1})) \right\}.$$

Thus:

$$\sum \left\{ a_k : m(w) < k \le g_i^q(m(w)) \right\} \ge$$
$$\sum \left\{ \sum \left\{ a_k : m(z) < k \le g_i(m(z)) \right\} : z \in C(w) \right\}.$$

Since the decomposition $\{z_r\}_{r=1}^q$ is admissible for w, we have:

$$q \le \min \left\{ (\min \ \text{supp} \ (z_r)) - 1 : r = 1, 2, \cdots, q \right\} \le m(w).$$

Thus, since $g_i^{m(w)}(m(w)) = g_{i+1}(m(w))$, we have:

$$\sum \{ a_k : m(w) < k \le g_{i+1}(m(w)) \}$$
$$\ge \ \sum \{ a_k : m(w) < k \le g_i^q(m(w)) \}$$
$$\ge \ \sum \{ \sum \{ a_k : m(z) < k, \le g_i(m(z)) \} \ z \in C(w) \},$$

and hence,

$$\sum \{ \|y\|_0 : y \in L_j \}$$
$$\le \ \sum \left\{ \sum \{ a_k : m(z) < k \le g_i(m(z)) \} : z \in P_j^i \right\}$$
$$\le \ \sum \left\{ \sum \{ \sum \{ a_k : m(z) < k \le g_i(m(z)) \} : z \in C(w) \} : w \in P_j^{i+1} \right\}$$
$$\le \ \sum \left\{ \sum \{ a_k : m(w) < k \le g_{i+1}(m(w)) \} : w \in P_j^{i+1} \right\}.$$

This proves (b), but unfortunately (a) may now fail. However, we can easily change the $m(w)$ terms to get (a) to hold, while keeping (b) and (c) true.

With this in mind, let $\{m(w_1), m(w_2), \cdots, m(w_q)\}$ be the ascending listing for $w_r \in P_j^{i+1}$. We leave $m(w_1)$ the same. (Note that this guarantees that after this re-selection, we still have (c).) If $m(w_2) < g_{i+1}(m(w_1))$, replace $m(w_2)$ by $g_{i+1}(m(w_1))$. Otherwise, leave it alone. It is clear that by continuing through the listing in this fashion that we can make (a) hold without changing the truth of (b) and (c). This completes the proof of (a), (b), and (c).

If we apply the claim when $i = j$, we get by conditions (iii) and (vi) of Lemma IV.c.2 that:

$$\sum \{ \|y\|_0 : y \in L_j \} \le$$
$$\sum \{ a_k : m(x) < k \le g_j(m(x)) \} \le$$
$$\sum \{ a_k : n < k \le g_j(n) \} = n^{j-1}.$$

Therefore the inequality of Proposition IV.c.3 holds when $k = \ell$, since $\|x\|_{\ell_1} = n$. $\quad \square$

Corollary IV.c.4: If $\ell, n \ge 2$, and I_n is the formal identity from $[t_k : n < k \le g_\ell(n)]$ into ℓ_1^m, (where $m := g_\ell(n) - n$,) then:

1. $\frac{2^{\ell+1}}{\ell+1} \le \|I_n\| \le 2^{\ell+1}$, and

2. $\lim_n \|I_n\| = 2^{\ell+1}$.

Proof:

(1) follows immediately from Propositions IV.b.1 and IV.c.3 by replacing each n by 2.

(2) follows from the proof of Proposition IV.c.3, since

$$\lim_n \sum_{j=1}^{\ell+1} 2^{-j} n^{j-\ell-1} = 2^{-\ell-1}.$$

$\square$

Finally we give:

Proof of Theorem IV.c.1: As noted after the statement of the theorem, we need prove only the "only if" part. So assume that $\{t_{k_n}\}_{n=1}^\infty$ is a subsequence of $\{t_n\}_{n=1}^\infty$ which is equivalent to $\{t_n\}_{n=1}^\infty$. Assume that there is no ℓ such that $k_n \leq g_\ell(n)$, for all large n, and let I_n denote the formal identity from $[t_i : k_n < i \leq k_{n+1}]$ onto $\ell_1^{m_n}$, (where $m_n := k_{n+1} - k_n$).

If $k_{n+1} \leq g_\ell(k_n)$, for all large n, then it is easily seen by induction on j that $k_{n+j} \leq g_\ell^j(k_n), \forall j$, and for all large n. Therefore,

$$
\begin{aligned}
g_{\ell+1}(k_n + j) \ &\geq \ g_\ell^{k_n+j}(k_n + j) \\
&\geq \ g_\ell^{k_n+j}(k_n) \qquad \geq k_{n+k_n+j} \\
&\qquad\qquad\qquad\qquad \geq k_{k_n+j}, \forall\, j, \text{ and all large } n.
\end{aligned}
$$

Hence, if $m := k_n + j$, $k_m \leq g_{\ell+1}(k_m)$, for all large m, contradicting our assumption that there is no ℓ with this property. Therefore, we may assume $k_n \leq g_\ell(k_n) \leq k_{n+1}$, infinitely often in both ℓ and n. But then, by Corollary IV.c.4, $\overline{\lim}_n \|I_n\| \geq \frac{2^\ell}{\ell+1}$, infinitely often in ℓ. Thus, $\overline{\lim}_n \|I_n\| = +\infty$.

Theorem IV.a.1 now implies that $\{t_n\}_{n=1}^\infty$ is not equivalent to $\{t_{k_n}\}_{n=1}^\infty$. $\square$

Notes and Remarks.

1. We cannot prove the results of this section in T^* by the same methods since we have relied heavily upon the analytic description of the norm on T. However, straightforward duality arguments show that all of the results of this section have corresponding statements in T^*.

2. The results of this section (and the preceding section) can be generalized to a subsequence $\{t_{k_n}\}_{n=1}^\infty$ of $\{t_n\}_{n=1}^\infty$. First, we generalize the "fast growing" hierarchy to $\{k_n\}_{n=1}^\infty$ by defining:

$$
\begin{cases}
g_0(k_n) := k_{n+1}, \text{ and} \\
g_{i+1}(k_n) := g_i^{k_n-1}(k_n), \ (i \geq 0).
\end{cases}
$$

Thus: $g_1(k_n) = k_{n+k_n}$, and $g_2(k_n)$ is a "stacked tower" obtained by continuing the following iteration k_n times:

$$k_{n+k_n},$$

$$k_{n+k_n+k_{n+k_n}},$$

$$k_{n+k_n+k_{n+k_n}+k_{n+k_n+k_{n+k_n}}}$$

.

.

.

Imitating the proofs of Lemma IV.c.2 and Proposition IV.c.3, we obtain:

Proposition IV.c.5: Let $\{t_{k_i}\}_{i=1}^\infty$ be a subsequence of $\{t_n\}_{n=1}^\infty$. and any $\forall \ell \geq 0, \exists x = \sum \{a_i t_{k_i} : n < i \leq g_{\ell+1}(k_n)\}$ such that:

$$\|x\| \leq \left(\sum_{j=1}^{\ell+1} 2^{-j} n^{j-\ell-1}\right) \sum \{|a_i| : n < i \leq g_{\ell+1}(k_n)\}.$$

In particular, since $n \geq 2$, replacing n by 2, we obtain:

Proposition IV.c.6: Let $\{t_{k_i}\}_{i=1}^\infty$ be a subsequence of $\{t_n\}_{n=1}^\infty$.
$\forall \ell \geq 0, \exists x = \sum_i a_i t_{k_i} \in T$ such that $\|x\|_{\ell_1} = 1$, and

$$2^{-\ell} \leq \|x\|_T \leq (\ell+1) \cdot 2^\ell.$$

We also have the theorem corresponding to IV.c.1, but we do not state it here. Instead, we should note the role that the number 2 plays in the results of this section, i.e., the occurences of 2^ℓ in Propositions IV.c.5 and IV.c.6 clearly come from the 2^{-1} in Construction I.1. In section X.B, we will define θ-Tsirelson's space, (denoted "T_θ"), for $0 < \theta < 1$, by replacing 2^{-1} by θ in Construction I.1. It's straightforward to check that Proposition IV.c.5 carries over to T_θ if we replace 2^{-1} by θ, and note that the requirement $n \geq 2$ becomes $n \geq \theta^{-1}$. We thus obtain:

Proposition IV.c.7: For all $\ell \geq 0$, and any n beyond $\theta^{-1}, \exists x = \sum \{a_i t_{k_i} : n < i \leq g_{\ell+1}(k_n)\}$, say, such that

$$\|x\| \leq \left(\sum_{j=1}^{\ell+1} \theta^j n^{j-\ell-1}\right) \|x\|_{\ell_1}.$$

What is particularly important about this Proposition is that it derives from Lemma IV.c.2 which has no reference to the number "2", (in our case: θ), i.e., the vector x given in Proposition IV.c.7 depends upon ℓ and n, but is independent of θ, as long as $n \geq \theta^{-1}$. Thus Proposition IV.c.6 becomes (in this new setting):

Proposition IV.c.8: Let $\{t_{k_i}\}_{i=1}^\infty$ be a subsequence of $\{t_n\}_{n=1}^\infty$. For each fixed $0 < \theta_0 < 1$, and $\ell \geq 1$, $\exists x = \sum_i a_i t_{k_i}$ such that $\|x\|_{\ell_1} = 1$, and $\forall \theta_0 \leq \theta < 1$, $\theta^{-\ell} \leq \|x\|_{T_\theta} \leq (\ell+1)\theta^{-\ell}$.

(Proposition IV.c.8 will be important in Chapter X for showing that T_{θ_1} and T_{θ_2} are totally incomparable, $\forall 0 < \theta_1 < \theta_2 < 1$.)

47

Chapter V: Modified Tsirelson's Space: T_M.

The purpose of this chapter is to develop a new tool for the study of Tsirelson's space T. We will show that T is naturally isomorphic to the so-called "modified" Tsirelson's space T_M defined by W. Johnson [27]. There are many new properties of T which can be approached through this isomorphism, since the definition of $\|\cdot\|_T$ implies statements about block basic sequences in T, while that of $\|\cdot\|_M$ implies statements about disjointly supported vectors in T_M. In particular, we will use this isomorphism to demonstrate that T enjoys the uniform projection property, but first we consider Johnson's construction.

Construction V.1:

As in Chapter I, let $\mathbb{R}^{(N)}$ be the (vector) space of all real sequences with finite support, and let $\{t_n\}_{n=1}^\infty$ denote the usual unit vector basis in $\mathbb{R}^{(N)}$. For $x = \sum_n a_n t_n \in \mathbb{R}^{(N)}$, we let:

$$
\left\{
\begin{array}{l}
\|x\|_{M,0} := \max_n |a_n|, \text{ and} \\[2mm]
\|x\|_{M,m+1} := \max\left\{\|x\|_{M,m}, \frac{1}{2}\max\left[\sum_{j=1}^k \|E_j x\|_{M,m}\right]\right\}, \ (m = 0,1,2,\cdots);
\end{array}
\right.
\tag{1}
$$

where the inner max in (1) is taken over all choices of disjoint finite subsets of natural numbers $\{E_j\}_{j=1}^k$ such that $k \leq E_j, (j = 1,2,\cdots,k), (k \in \mathbb{N})$. (Any expression of form $\frac{1}{2}\sum_{j=1}^k \|E_j x\|_{M,m}$ will be called an <u>allowable sum for x</u>. Then $\|x\|_M := \lim_m \|x\|_{M,m}$ defines a norm on $\mathbb{R}^{(N)}$ with a unique extension (also denoted $\|\cdot\|_M$) to the completion T_M of $\mathbb{R}^{(N)}$. T_M is <u>modified Tsirelson's spac</u>

Since there are in general many more allowable sums for a vector $x \in \mathbb{R}^{(N)}$ than admissible sums, it was surprising when in [21] it was proved that these two notions lead to equivalent norms. To simplify the proof of this, we will first state a lemma (actually, a special case of the theorem).

To simplify the statement of the lemma, we introduce a new norm $\||\cdot\||$ on T which is defined exactly the same way as the regular norm on T with the change that for any $x \in T$, an admissible sum for x with respect to $\||\cdot\||$ is a sum of form $\frac{1}{2}\sum_{j=1}^{2k} \||E_j x\||$, where $k \leq E_1 < E_2 < \cdots < E_{2k}$. By remark (3) of the "Notes and Remarks" of Chapter I, we have:

$$
\|x\| \leq \||x\|| \leq 3\|x\|, \ (x \in T).
\tag{2}
$$

We can now state the lemma and the main theorem of this chapter.

Lemma V.2: For $m = 0,1,2,\cdots$, and $k_0 \in \mathbb{N}$, and any disjointly supported vectors $\{u_i\}_{i=1}^s$ in $[t_{2^n}]_{n=k_0}^\infty$ with $s \leq 2^{k_0}$, the following holds:

$$
\sum_{i=1}^s \|u_i\|_{M,m} \leq 2\||\sum_{i=1}^s u_i\||.
\tag{3}
$$

Before we prove Lemma V.2, let's use it to prove:

Theorem V.3: The unit vector bases of T and T_M are equivalent, i.e., the formal identity between T and T_M is an isomorphism.

Proof: By Lemma V.2, for any $x = \sum_n a_n t_{2^n} \in \mathbb{R}^{(N)}$, we have:

$$\|x\|_M \leq \|\|x\|\|. \tag{4}$$

By inequality (2), $\|\|x\|\| \leq 3\|x\|$, $\forall x \in \mathbb{R}^{(N)}$.

By Corollary IV.b.2, $\exists A$ such that:

$$\|\sum_n a_n t_{2^n}\| \leq A\|\sum_n a_n t_n\|, \text{ for all scalers } \{a_n\}_{n=1}^{\infty}.$$

Putting this all together, for all sequences of scalars $\{a_n\}_{n=1}^{\infty}$ which are eventually zero,

$$\|\sum_n a_n t_{2^n}\| \leq M\|\|\sum_n a_n t_{2^n}\|\|$$

$$\leq 3\|\sum_n a_n t_{2^n}\| \leq 3A\|\sum_n a_n t_n\|$$

$$\leq 3A\|\sum_n a_n t_n\|_M \leq 3A\|\sum_n a_n t_{2^n}\|_M.$$

This implies that $\|\cdot\|$ and $\|\cdot\|_M$ are equivalent. $\square$

Proof of Lemma V.2:

We proceed by induction on m, with the case $m = 0$ obviously true. So assume that the lemma holds for some m. Without loss of generality we may assume that the vectors u_i are finitely supported. It's easy to believe (and not difficult to prove) that Proposition I.10 holds for $\|\cdot\|_M$ as well as $\|\cdot\|_T$. Hence the vectors u_i may be re-indexed so that for some $1 \leq s_1 \leq s$, we have:

$$\|u_i\|_{M,m+1} = \|u_i\|_{M,0} \text{ iff } s_1 < i \leq s, \text{ and} \tag{5}$$

$$\min \text{ supp } (u_i) < \min \text{ supp } (u_{i+1}), \text{ for } 1 \leq i \leq s_1. \tag{6}$$

Now we can select $k_0 \leq k(1) < k(2) < \cdots < k(s_1)$ and pairwise disjoint (or void) subsets $\{E_j^i\}_{j=1}^{2^{k(i)}}$ of $\{2^{k(i)}, 2^{k(i)+1}, \ldots\}$ such that

$$2^{k(i)} = \min \text{ supp } (u_i), \text{ and}$$

$$\|u_i\|_{M,m+1} = \frac{1}{2} \sum \{\|E_j^i u_i\|_{M,m} : 1 \leq j \leq 2^{k(i)}\}.$$

Now choose $k(s_1 + 1) \in \mathbb{N}$ so that $k(s_1 + 1) \geq E_j^i, \forall i, j$, and let Q_n denote the natural projection of T onto $[t_i]_{i=1}^n$, $(n = 1, 2, \cdots)$. (To simplify notation below, we denote by $\{t_n\}_{n=1}^{\infty}$ the unit vector basis of T and T_M subscripting the norm to distinguish the two usages.)

Since $Q_{2^{k(i)-1}} E_j^i u_i = 0$, $\forall i, j$, the triangle inequality yields:

$$\sum_{i=1}^{s_1} \|U_i\|_{M,m+1}$$

$$= \sum_{i=1}^{s_1} \left[\frac{1}{2} \sum_{j=1}^{s^{k(i)}} \|E_j^i u_i\|_{M,m} \right]$$

$$\leq \frac{1}{2} \sum_{i=1}^{s_1} \sum_{j=1}^{s^{k(i)}} \sum_{n=i}^{s_1} \|(Q_{2^{k(n+1)-1}} - Q_{2^{k(n)-1}}) E_j^i u_j\|_{M,m}.$$

Since $2^{k(\ell)} \in \text{supp} \, (u_i)$ iff $\ell = i$, we may continue this inequality:

$$\leq \frac{1}{2} \sum_{i=1}^{s_1} |a_{2^{k(i)}}| + \frac{1}{2} \sum_{i=1}^{s_1} \sum_{j=1}^{2^{k(i)}} \sum_{n=i}^{s_1} \| (Q_{2^{k(n+1)-1}} - Q_{2^{k(n)}}) E_j^i u_i \|_{M,m}$$

$$= \frac{1}{2} \sum_{i=1}^{s_1} |a_{2^{k(i)}}| + \frac{1}{2} \sum_{n=1}^{s_1} \left[\sum_{i=1}^{n} \sum_{j=1}^{2^{k(i)}} \| (Q_{2^{k(n+1)-1}} - Q_{2^{k(n)}}) E_j^i u_i \|_{M,m} \right].$$

Examining this final bracketed expression (for fixed n), we see that it consists of $\sum_{i=1}^{n} 2^{k(i)} \leq 2^{k(n)+1} \leq 2^{k(n+1)}$ disjointly supported vectors in

$$[t_{2j} : k(n+1) \leq j].$$

Applying the induction hypothesis to each such term yields:

$$\sum_{i=1}^{n} \sum_{j=1}^{2^{k(i)}} \| (Q_{2^{k(n+1)-1}} - Q_{2^{k(n)}}) E_j^i u_i \|_{M,m}$$

$$\leq 2 \| \sum_{i=1}^{n} \sum_{j=1}^{2^{k(i)}} (Q_{2^{k(n+1)-1}} - Q_{2^{k(n)}}) E_j^i u_i \|$$

$$= 2 \| (Q_{2^{k(n+1)-1}} - Q_{2^{k(n)}}) \left(\sum_{i=1}^{s_1} u_i \right) \|.$$

For all $s_1 < i \leq s$, choose $q(i)$ so that $\|u_i\|_{M,m+1} = |a_{2^{q(i)}}|$. Note that the sets $\left\{ 2^{k(i)} : 1 \leq i \leq s_1 \right\}$ and $\left\{ 2^{q(i)} : s_1 < i \leq s \right\}$ are disjoint, and reindex their union as $\left\{ 2^{m(i)} : 1 \leq i \leq s \right\}$, where $m(1) < m(2) < \cdots < m(s)$. We now have

$$\sum_{i=1}^{s} \|u_i\|_{M,m+1} = \left(\sum_{i=1}^{s_1} + \sum_{i=s_1+1}^{s} \right) \|u_i\|_{M,m+1}$$

$$\leq \frac{1}{2} \sum_{i=1}^{s_1} |a_{2^{k(i)}}| + \sum_{n=1}^{s_1} 2 \| (Q_{2^{k(n+1)-1}} - Q_{2^{k(n)}}) \left(\sum_{i=1}^{s_1} u_i \right) \| + \frac{1}{2} \sum_{i=s_1+1}^{s} |a_{2^{q(i)}}|$$

$$\leq \sum_{i=1}^{s} \| a_{2^{m(i)}} t_{2^{m(i)}} \| + \sum_{n=1}^{s_1} \| (Q_{2^{m(n+1)-1}} - Q_{2^{m(n)}}) \left(\sum_{i=1}^{s_1} u_i \right) \|$$

$$\leq 2 \| \sum_{i=1}^{s} u_i \|.$$

This last inequality holds because the right side of the previous line consists of no more than $2s$ disjoint blocks of $\sum_{i=1}^{s} u_i$ supported in $\{t_i\}_{i=s}^{\infty}$, and hence is an admissible sum for $\| \cdot \|$. This completes the induction and the proof of the lemma. $\square$

We now list some of the consequences of the isomorphism between T and T_M (listing others in the notes at chapter's end). First a few definitions:

Definition V.4:

a). A Banach space X is said to have the uniform approximation property, (U.A.P.), if $\exists \lambda > 0$ and a function $n(k)$ such that whenever E is a k-dimensional subspace of X, $\exists$ an operator $T : X \to X$ for which:

    i). $T|_E$ is the identity,

ii). $\|T\| \leq \lambda$, and

iii). $\dim TE \leq n(k)$.

b). If (in addition to the above) T can be chosen to be a projection, then X has the uniform pro jection property ("U.P.P.").

We will need here (and later on) a special property of the unit vector basis of ℓ_1 which is given by the following:

Remark V.5: The unit vector basis of ℓ_1 is block injective [41], i.e., whenever the unit vector basis of ℓ_1 is represented as a disjointly supported sequence with respect to any unconditional basis of a Banach space X, then it spans a complemented subspace of X. It follows that $\exists$ a function $\lambda(K)$ such that for each n whenever $\{y_i\}_{i=1}^n$ is K -equivalent to the unit vector basis of ℓ_1^n and is disjointly supported with respect to a 1-unconditional basis $\{x_i\}_{i=1}^\infty$ for X, then $[y_i]_{i=1}^n$ is $\lambda(K)$ -complemented in X.

We now show that T_M (and hence T also) has U.P.P. We need an argument due to Kwapien [46]. The form given here was first observed by W. Johnson and presented in [42]. (We omit the proof.)

Proposition V.6: There is a function $N(k, \epsilon)$ (in fact, $N(k, \epsilon) := \left[\left[\frac{2k^2}{\epsilon}\right]\right]^k$) such that for any fixed $\epsilon > 0$, every Banach lattice L, and every k -dimensional subspace F of L, there are $N := N(k, \epsilon)$ disjoint elements $\{g_j\}_{j=1}^N$ in L and a linear operator

$$V : F \to G := [g_j]_{j=1}^N$$

such that

$$\|Vf - f\| \leq \epsilon\|f\|, \ \forall f \in F.$$

We observe the following:

Proposition V.7: The space T_M (and hence also T) has U.P.P.

Proof: Let E be a k -dimensional subspace of T_M, for some $k \in \mathbb{N}$. By Proposition V.6, without loss of generality we may assume that $\exists$ vectors $\{x_i\}_{i=1}^k$ disjointly supported in T_M for which $E = [x_i]_{i=1}^k$. Let P_k be the natural projection of T_M onto $[t_i]_{i=1}^k$, and let $I := \{i : 1 \leq i \leq k$, and $(I - P_k)x_i \neq 0\}$. By the definition of norm in T_M, the sequence $\left\{\frac{(I-P_k)x_i}{\|(I-P_k)x_i\|}\right\}_{i \in I}$ is 2-equivalent to the unit vector basis of $\ell_1^{\bar{I}}$ (i.e., n disjointly supported vectors beyond n are 2-equivalent to a ℓ_1 -sequence). Since the unit vector basis of ℓ_1 is block injective, $\exists \lambda := \lambda(2) > 0$ and a projection Q of $[t_i : i > k]$ onto $\left[\frac{(I-P_k)x_i}{\|(I-P_k)x_i\|} : i \in I\right]$ with $\|Q\| \leq \lambda$.

It follows that the operator $P : T_M \to T_M$ defined by

$$Px := P_k x + Q(O - P_k)x$$

is a projection of T_M onto a subspace F of T_M with $E \subset F$. Since $\|P\| \leq 2\lambda$, and $\dim F \leq 2k$, the proof is complete. $\square$

Another property which is obvious for T_M (and now holds in T) is given in:

Proposition V.8: $\exists$ a universal constant $K > 0$ such that any disjointly supported vectors $\{y_i\}_{i=1}^n$ in $[t_i]_{i=n}^\infty$ are K -equivalent to the unit vector basis of ℓ_1^n. Therefore, if $\{y_i\}_{i=1}^n$ are disjointly supported vectors in T, then $\exists$ a subset $E :\subset \{1, 2, \cdots, n\}$ with $\overline{\overline{E}} \geq \frac{n}{2}$ and so that $\{y_i : i \in E\}$ is K -equivalent to the unit vector basis of $\ell_1^{\overline{\overline{E}}}$.

Finally we note that obvious modifications of the proofs of Lemmas II.1 and II.3 yield:

Proposition V.9: Let $\{E_n\}_{n=1}^\infty$ be a sequence of finite pairwise disjoint subsets of $\mathbb{N}$. Define

$$i(n) := \min E_n, \text{ and } j(n) := \max E_n, (n \in \mathbb{N}),$$

and let $y_n = \sum \{a_i t_i : i \in E_n\}$ be normalized vectors in T. Then for all choices of scalars $\{b_n\}_{n=1}^\infty$, we have:

$$\left\| \sum_n b_n t_{i(n)} \right\| \leq \left\| \sum_n b_n y_n \right\|, \text{ and} \tag{a}$$

$$\left\| \sum_n b_n y_n \right\| \leq K \left\| \sum_n b_n t_{j(n)} \right\|, \tag{b}$$

where K is a universal constant.

Notes and Remarks:

1/. All of the results in this chapter have obvious dualizations in T^*.

2/. Tsirelson's space T is the first space besides c_0 and the ℓ_p -spaces $(1 \leq p < \infty)$ in which the uniform projection property is trivial.

3/. Proposition V.9 is a generalization of Lemmas II.1 and II.3. These latter lemmas were used to prove that the closed span of every bounded block basic sequence in T is complemented and that every bounded block basic sequence in T is equivalent to a subsequence of $\{t_n\}_{n=1}^\infty$. The corresponding conclusion from Proposition V.9 cannot be drawn because of a result of J. Lindenstrauss and L. Tzafriri [37] which implies that T must possess disjointly supported sequences of vectors which span uncomplemented subspaces of T. The problem is that an important ingredient in the proof of Proposition II.4 which is needed here is for $\{t_{i(n)}\}_{n=1}^\infty$ to be equivalent to $\{t_{j(n)}\}_{n=1}^\infty$, but this is not true in general, so $\{y_n\}_{n=1}^\infty$ need not be equivalent to a subsequence of $\{t_n\}_{n=1}^\infty$.

4/. If $p := \inf \{q > 1 : T \text{ has lower } q\text{- estimate }\}$, then a result [5] of G. Pisier implies that the unit vector bases of $\ell_p^n, (n \in \mathbb{N})$, are uniformly representable as disjointly supported vectors in T. By Proposition V.8, this is impossible, unless $p = 1$. Thus we have:

Proposition V.10: Tsirelson's space T has a lower q -estimate for all $q > 1$.

(It follows that T^* has an upper p -estimate for all $p > 1$.)

5/. A stronger result than Proposition V.9 is proved in Chapter IX. There we drop the assumption that the $\{y_n\}_{n=1}^\infty$ are disjointly supported and insist instead upon only a restriction on the number of vectors y_n which may have non-trivial support in $\{1, 2, \cdots k\}$, for any k.

6/. The proof of Corollary IV.b.2 can be adapted to the space T_M to give the stronger result:

Proposition V.11: If $i \geq 1$, and $\{y_j : n \leq j \leq g_i(n)\}$ are disjointly supported and normalized in $[t_j]_{j=n}^\infty \subset T_M$, then $\{y_j : n \leq j \leq g_i(n)\}$ is 2^{i+1} -equivalent to the unit vector basis of ℓ_1^m (where $m := g_i(n) - n + 1$, and the g_i are defined as in Chapter IV.B.).

As an immediate consequence of this, we have:

Proposition V.12: For any n and $i \geq 0$, let $I_n, I_{n+1}, \cdots I_{g_i(n)}$ be a partition of $\{n, n+1, \cdots\}$. If $X_j := [t_k : k \in I_j]$, $(j = n, n+1, \cdots, g_i(n))$, then,

$$\|I\| \cdot \|I^{-1}\| \leq 2^i,$$

where I is the formal identity map from

$$[t_k]_{k=n}^\infty \text{ to } \left(\sum \oplus \{X_j : n \leq j \leq g_i(n)\}\right)_{\ell_1^{m_n}},$$

(where $m_n := g_i(n) - n + 1$.)

7/. The space ℓ_1 has the following property [35]: there is a function $f(\cdot, \cdot)$ so that for every $\|P\|$ -complemented subspace X of ℓ_1 with a K -unconditional basis $\{x_i\}_{i=1}^n$, the sequence $\{x_i\}_{i=1}^n$ is $f(K, \|P\|)$ -equivalent to the unit vector basis of ℓ_1^n. This (combined with the proof of Proposition V.7) yields:

Proposition V.13: $\exists$ a function $f(\cdot, \cdot)$ so that if $\{x_i\}_{i=1}^n$ is a K -unconditional basis for a $\|P\|$ -complemented subspace of $[t_i]_{i=n}^\infty$, then $\{x_i\}_{i=1}^n$ is $f(K, \|P\|)$ -equivalent to the unit vector basis of ℓ_1^n.

(We will omit a formal proof of this, since Proposition VII.a.3 contains this result as a special case.)

8/. In Chapter I, we saw that ℓ_1^n was 2-equivalent to a subspace of $[t_k]_{k=n}^\infty$ in T. In T_m, we have the analogous result:

Proposition V.14: If $\{u_i\}_{i=1}^n$ are normalized disjointly supported blocks against $\{t_n, t_{n+1}, \cdots\}$ in T_M, then $[u_i]_{i=1}^n$ is 2-equivalent to ℓ_1^n.

(We used this in the proof of Proposition V.7.)

9/. Proposition II.4 can be reformulated and proven for T_M (with different embedding constants).

Chapter VI: Embedding Theorems about T and T*

In 1960, A. Pelczynski [45] showed that for c_0 and the spaces ℓ_p, $(1 \leq p < \infty)$ any infinite-dimensional subspace necessarily contains an isomorphic copy of the original space. H. Rosenthal accordingly introduced the notion of "minimality".

Definition VI.1: An infinite-dimensional Banach space is <u>minimal</u> if it embeds into each of its infinite-dimensional subspaces.

H. Rosenthal went on to ask whether every infinite-dimensional Banach space contained a minimal subspace. Here we answer this question by demonstrating that T contains no minimal subspaces, and (for good measure) we add T^* to Pelczynski's list of minimal Banach spaces.

A. The minimality of T*.

In this section we present some results of P. Casazza, W. Johnson, and L. Tzafriri [17] which imply that T^* is a minimal Banach space.

Theorem VI.a.1: Any infinite-dimensional subspace of a quotient space of T^* contains a subspace isomorphic to T_*.

Proof: Let X be any infinite-dimensional subspace of a quotient space of T^*, and let X_0 be a subspace of X with a Schauder basis $\{x_n\}_{n=1}^{\infty}$. By interpreting this basis as an F.D.D. for X_0 and applying the Blocking Principle (Theorem III.6), we conclude that $\{x_n\}_{n=1}^{\infty}$ contains a subsequence which is equivalent to a subsequence $\{t_{k_n}^*\}_{n=1}^{\infty}$ of $\{t_n^*\}_{n=1}^{\infty}$ (where $\{t_n^*\}_{n=1}^{\infty}$ is the canonical unit vector basis of T^*.) Therefore in order to complete the proof it suffices to show that for any increasing sequence $\{k_n\}_{n=1}^{\infty}$ of natural numbers, T^* is isomorphic to a subspace of $[t_{k_n}^*]_{n=1}^{\infty}$.

Toward this end, let $\{k_n\}_{n=1}^{\infty}$ be such a sequence, and let $V := [t_{k_n}^*]_{n=1}^{\infty}$.

Choose $m_1 = 1$, and let

$$\begin{cases} E_1 := \{1, 2, \cdots, k_{m_1}\}, \text{ and} \\ U_1 := [t_n^* : n \in E_1]. \end{cases}$$

Since $\left\{ t_{k_n}^* : h < n \leq 2h \right\}$ is 2-equivalent to the unit vector basis of ℓ_{∞}^h, $\forall\ h = 1, 2, \cdots$, and since ℓ_{∞} is a universal space for all separable spaces, we may choose an n_1 such that for

$$\begin{cases} F_1 := \{k_1, k_2, \cdots, k_{n_1}\}, \text{ and} \\ V_1 := [t_n^* : n \in F_1], \end{cases}$$

there exists an invertible operator $L_1 : U_1 \to V_1$ with

$$\|L_1^{-1}\| = 1, \text{ and } \|L_1\| \leq 2.$$

Now let

$$\begin{cases} m_2 := \max\{m_1, n_1\} + 1, \text{ and} \\ E_2 := \{k_{m_1} + 1, k_{m_1} + 2, \cdots, k_{m_2}\}, \text{ and} \\ U_2 := [t_n^* : n \in E_2]. \end{cases}$$

54

As before, choose n_2 so that for

$$\begin{cases} F_2 := \{k_{n_1+1}, k_{n_1+2}, \cdots, k_{n_2}\}, \text{ and} \\ V_2 := [t_n^* : n \in F_2], \end{cases}$$

there exists an invertible operator $L_2 : U_2 \to V_2$ with

$$\|L_2^{-1}\| = 1, \text{ and } \|L_2\| \le 2.$$

Continuing in this fashion, we construct:

a). sequences of finite subsets of $\mathbb{N}$, $\{E_j\}_{j=1}^\infty$ and $\{F_j\}_{j=1}^\infty$,

b). subspaces of T^*,
$$U_j := [t_n^* : n \in E_j], \quad (j \in \mathbb{N}), \text{ and}$$
$$V_j := [t_n^* : n \in F_j], \quad (j \in \mathbb{N}), \text{ and}$$

c). invertible operators $L_j : U_j \to V_j$, $(j \in \mathbb{N})$, such that

(i). $1 \le E_1 < E_2 < \cdots$, and $1 \le F_1 < F_2 < \cdots$,

(ii). $\bigcup_j E_j = \mathbb{N}$, and $\bigcup_j E_j = \{k_1, k_2, \cdots\}$,

(iii). $F_j \subset E_j \cup E_{j+1}$, $(j \in \mathbb{N})$, and

(iv). $\|L_j^{-1}\| = 1$, and $\|L_j\| \le 2$, $(j \in \mathbb{N})$.

Now by Theorem III.5 (and the discussion following it) we can conclude that the operator $L : T^* \to V$, defined by

$$Lx = \sum_j L_j(E_j x), \quad (x \in T^*),$$

is an isomorphism from T^* onto a subspace of V. $\square$

Our next proposition is from [17] and shows that F.D.D.'s of type T and of type T^* have one of the important properties of ℓ_p-sums of finite-dimensional spaces, for $1 < p < \infty$.

Proposition VI.a.2: Let X be a Banach space having an F.D.D. of type T (respectively: of type T^*). Then every quotient space of X is isomorphic to a subspace of a space with an F.D.D. of type T (respectively: of type T^*).

Proof: Let X be a Banach space with an F.D.D. of type T (respectively: of type T^*). Let V be a quotient space of X. If V has an F.D.D., then this Proposition is a special case of the Blocking Principle (Theorem III.6). The general case follows with only notational changes from the proof of the corresponding statement for ℓ_p-sums. (see Theorem 1 of [30].) $\square$

A well known result of W. Johnson and M. Zippin [30] asserts that quotient spaces of c_0 are isomorphic to subspaces of c_0. D. Alspach [3] strengthened this to show that quotient spaces of c_0 are almost isometric to subspaces of c_0. As an application of these results, we prove a similar assertion for T^* from [17].

Theorem VI.a.3: Every quotient space of T^* is isomorphic to a subspace of T^*. (Hence, every subspace of T is isomorphic to a quotient of T.)

Proof: Let X be a quotient space of T^*. By Proposition VI.a.2, X is isomorphic to a subspace of a space Y which has an F.D.D. of type T^*. To complete the proof, it suffices to show that Y embeds isomorphically into T^*. But this is clearly the case (by Theorem III.7). $\quad\Box$

Notes and Remarks:

1/. Besides the subspaces of c_0 and ℓ_p, $(1 \leq p < \infty), T^*$ is the only known example of a minimal Banach space. A. Pelczynski [45] actually showed that if X is c_0 or some ℓ_p, $(1 \leq p < \infty)$, then every infinite-dimensional subspace of X contains an isomorph of X which is complemented in X. If this property is termed complemented minimality, then it can be shown that T^* is minimal, but not complemented minimal. (Indeed, the complemented minimality of T^* would imply the minimality of T, which is not even close to being true.)

2/. The properties of Proposition VI.a.2 were known to hold only in c_0 -sums and ℓ_p -sums $(1 \leq p < \infty)$ of finite-dimensional spaces prior to this proposition. Theorem VI.a.3 was previously known to hold only in c_0.

3/. The proofs of Theorems VI.a.1 and VI.a.3 use the universality of ℓ_∞ for separable spaces and the fact that ℓ_∞^n embeds uniformly (and hence uniformly complementably) into $[t_i^*]_{i=m}^\infty$, $\forall\, n, m = 1, 2, \cdots$. We will see in the next section that these theorems fail in T.

4/. It of course follows that every infinite-dimensional subspace of T^* is minimal, since every infinite-dimensional subspace of a minimal space must be minimal.

B. The non-minimality of T.

We show here that T contains no minimal subspaces and that T is not a primary Banach space, i.e., $\exists$ Banach spaces X and Y such that $T \approx X \oplus Y$, while $X \not\approx T \not\approx Y$. First we recall a definition:

Definition VI.b.1: A Banach space Y is crudely finitely representable in a Banach space X if $\exists$ a constant C so that for every finite-dimensional subspace F of Y, $\exists$ a subspace W of X and an invertible operator $L : F \to W$ such that $\|L\| \cdot \|L^{-1}\| \leq C$.

Clearly, if Y embeds isomorphically into X, then Y is crudely finitely representable in X. (The converse fails, since every Banach space is crudely finitely representable in c_0.)

H. Rosenthal has shown [48] that a subspace of L_1 which contains the spaces ℓ_1^n uniformly must also contain ℓ_1, while J. Lindenstrauss and A. Pelczynski [34] have shown that any separable Banach space which is crudely finitely representable in L_1 must embed in ℓ_1. Our first proposition follows immediately from these facts.

Proposition VI.b.2: Tsirelson's space T is not crudely finitely representable in L_1.

Our next proposition is due to P. Casazza and E. Odell [21,23], and is a "local" answer to the embedding problem of T into its subspaces.

Proposition VI.b.3: For any natural numbers $k(1) < k(2) < \cdots$, and $n \geq 1$, and any $K > 0, \exists m \in \mathbb{N}$ such that every operator

$$L : \left[t_{k(i)} : n \leq i \leq m\right] \to \left[t_{k(i)} : 1 \leq i \leq n, \text{ or } i > m\right]$$

satisfies

$$\|L\| \cdot \|L^{-1}\| \geq K.$$

Proof: (It suffices to prove this result in T_M .)

Fix $n \in \mathbb{N}, k(1) < k(2) < \cdots$, and $K > 0$. For each $s \geq n$, let

$$a_s := \inf\left\{d\left(\left[t_{k(i)} : n \leq i \leq s\right], Z\right) : Z \text{ is a subspace of } L_1\right\},$$

where d is the Banach-Mazur distance. By Proposition VI.b.2, we have $\lim_s a_s = +\infty$.

Now choose $M > 4$ so that

(1). $\sup\left\{d(W, Z) : Z \text{ a subspace of } L_1, W \text{ a subspace of } \left[t_{k(i)} : 1 \leq i < n\right]\right\} \leq M$.

Now choose m beyond n so large that $\frac{2m}{2^5 \cdot M} > K$.

$$\text{Let} \begin{cases} W_1 := \left[t_{k(i)} : 1 \leq i < n\right], \\ W_2 := \left[t_{k(i)} : n \leq i \leq m\right], \text{ and} \\ W_3 := \left[t_{k(i)} : i > m\right], \end{cases}$$

and let L be any operator such that

$$L : W_2 \to W_1 \oplus W_3 = \left[t_{k(i)} : 1 \leq i < n, \text{ or } i > m\right].$$

By Proposition V.6, $L(W_2)$ is 2-isomorphic to a subspace of $W_1 \oplus W_3$ which is spanned by $[4(m-n)^2]^{m-n}$ disjointly supported vectors.

Let $p := [4(m-n)^2]^{m-n}$ and let $\{y_i\}_{i=1}^p$ be the disjointly supported vectors in $W_1 \oplus W_3$ for which $[y_i : 1 \leq i \leq p]$ contains a subspace 2-isomorphic to $L(W_2)$.

$$\text{Let} \begin{cases} I := \left\{1 \leq i \leq p : P_{k(m)} y_i \neq 0\right\}, \\[1em] J := \left\{1 \leq i \leq p : (I - P_{k(m)}) y_i \neq 0\right\}, \\[1em] A := \left[P_{k(m)} y_i : i \in I\right], \text{ and} \\[1em] B := \left[P_{k(m)} y_i : i \in J\right], \end{cases}$$

where the $P_{k(m)}$ are the obvious projections.

Since $\bar{\bar{J}} \leq p$, and supp $(I - P_{k(m)}) y_i \geq k(m+1)$, for all $i \in J$, it follows by Proposition IV.b.1 and Corollary IV.b.2 (since $p \leq g_3(m)$) that $\left\{\left(I - P_{k(m)}\right) y_i : i \in J\right\}$ is 2^4-equivalent to the unit vector basis of $\ell_1^{\bar{\bar{J}}}$.

That is:

(2). $\inf\{d(B, W) : W$ is a subspace of $L_1\} \leq 2^4$. By (1) above, $\exists$ a subspace W of L_1 such that

(3). $d(A < W) \leq M$. By (2), (3), and the fact that $[y_i]_{i=1}^p \subset A \oplus B$, we have:

(4). $\inf\{d([y_i]_{i=1}^p, W) : W$ is a subspace of $L_1\} \leq 2^4 \cdot M$.

Since $\left[Lt_{k(i)}\right]_{i=n}^m$ is 2-isomorphic to a subspace of $[y_i]_{i=1}^p$, we see that

$$a_m = \inf\left\{d\left(\left[t_{k(i)}\right]_{i=n}^m, W\right) : W \text{ is a subspace of } L_1\right\}$$
$$\leq 2^5 \cdot M \cdot \|L\| \cdot \|L^{-1}\|.$$

Thus by our choice of m,

$$\|L\| \cdot \|L^{-1}\| \geq \frac{a_m}{2^5 M} > K.$$

$\square$

As an application of this result, we prove a theorem which yields all of the results stated at the beginning of this section.

Theorem VI.b.4: Let $\{t_{k_n}\}_{n=1}^\infty$ be a subsequence of $\{t_n\}_{n=1}^\infty$ Then $\exists$ a partition $\{I, J\}$ of the natural numbers such that if

$$X := [t_{k_n} : n \in I], \text{ and } Y := [t_{k_n} : n \in J],$$

then neither X nor Y is crudely finitely representable in the other.

Proof: By Proposition VI.b.3, we may choose natural numbers $n_1 < n_2 < \cdots$ so that every operator

$$L : [t_{k_i} : n_j < i \leq n_{j+1}] \rightarrow [t_{k_i} : 1 \leq i \leq n_j, \text{ or } i > n_{j+1}]$$

satisfies $\|L\| \cdot \|L^{-1}\| \geq j$, for $j = 1, 2, \cdots$.

Thus if

$$\begin{cases} X := [t_{k_i} : n_{2j} < i \leq n_{2j+1}; & j = 1, 2, \cdots], \text{ and} \\ Y := [t_{k_i} : n_{2j-1} < i \leq n_{2j}; & j = 1, 2, \cdots], \end{cases}$$

then neither X nor Y is crudely finitely representable in the other. $\square$

Corollary VI.b.5: Tsirelson's space T is not primary.

Proof: If $k_n := n$ in the above theorem, then $T \approx X \oplus Y$, (X, Y defined above), and T is crudely finitely representable neither in X nor in Y. $\square$

Corollary VI.b.6: Tsirelson's space T has no minimal subspaces.

Proof: If X is an infinite-dimensional subspace of T, then by Proposition II.7, X must have a complemented subspace Y isomorphic to $[t_{k_n}]_{n=1}^\infty$, for some $k_1 < k_2 < \cdots$. By the discussion following Proposition I.14, we have that $Y \approx Y \oplus Y$, and hence that $X \approx X \oplus Y$. By Theorem VI.b.4, $Y \approx Y_1 \oplus Y_2$, say, where neither Y_1 nor Y_2 is crudely finitely representable in the other. Hence X is not crudely finitely representable in its subspace Y_1, and thus X is not minimal. $\square$

Notes and Remarks:

1/. Although there are subspaces of quotient spaces of T which do not contain isomorphs of T, a portion of the proofs in section A for T^* carry over to T. In particular, every infinite-dimensional subspace of a quotient space of T contains a subspace isomorphic to $[t_{k_n}]_{n=1}^\infty$, for some $k_1 < k_2 < \cdots$.

2/. It's also clear that T^* is not primary, since this would force T to be primary also. Although Proposition VI.b.3 fails in T^* as stated, its conclusion does hold under the additional hypothesis that the ranges of the operators L are uniformly complemented in T^*. This result is contained in Proposition VII.a.4, and yields a strengthening of Proposition VI.b.3.

3/. Our best guess is that neither T nor T^* has any primary subspaces, but we're at a loss for a proof.

4/. The following "Schroeder-Bernstein"-type problems are also open:

Problem VI.b.7: If X is a Banach space and X and T are isomorphic to complemented subspaces of each other, must X and T be isomorphic?

Problem VI.b8.: If X is a subspace of T, and if T embeds into X, then must T be isomorphic to a complemented subspace of X?

(In other words: does every subspace of T which is isomorphic to T contain a complemented subspace which is isomorphic to T?)

5/. R. C. James has shown that both c_0 and ℓ_1 embed almost isometrically into each super-space. A classical problem is whether ℓ_p, $(1 < p < \infty)$ satisfies the same type of "distortion" theorem. Now all of these spaces are minimal, so we are led to query:

Problem VI.a.9: Does T^* satisfy a "distortion" theorem?

Chapter VII: Isomorphisms between subspaces of Tsirelson's space which are spanned by subsequences of $\{t_n\}_{n=1}^{\infty}$.

Our main result in this chapter is a proof of the assertion that two subsequences of $\{t_n\}_{n=1}^{\infty}$ span isomorphic subspaces of T iff the subsequences are equivalent. The main tool used for this result is a quantitative version of Theorem VI.b.3. First we prove several technical lemmata which have broad implications in the study of the structure of isomorphisms on T.

A. Technical Lemmata.

Recall the following from Chapter IV:

for all $n \in \mathbb{N}$, $\begin{cases} \exp_0(n) : & = n, \text{and} \\ \exp_{m+1}(n) : & = 2^{\exp_m(n)}, \text{for } m \geq 0 \end{cases}$.

Lemma VII.a.1: Let $Z = X \oplus Y$ be the direct sum of two finite-dimensional Banach spaces X and Y, and let P be a projection from Z onto a finite-dimensional subspace. Let $\{z_i : z_i = x_i + y_i, x_i \in X, y_i \in Y, i = 1, 2, \cdots, k\}$ be a normalized K-unconditional basis for PZ. Let $z_1^*, z_2^*, \cdots, z_k^*$ be chosen from Z^* so that

$$Pz = \sum_{i=1}^{k} z_i^*(z)z_i, \quad \forall z \in Z \ .$$

Then for all $0 < c < 1, \exists$ an integer j $(j := j(\|P\|, c))$ such that if

$$A := \{i : |z_i^*(x_i)| \geq c\},$$

then

$$\bar{A} \leq \exp_j(\dim X).$$

Proof: Choose $\epsilon > 0$ small enough so that both $\dfrac{c - K\epsilon}{K}$ and $\dfrac{c}{\|P\|} - \epsilon$ are positive, and then choose a natural number m such that:

$$m\left(\frac{c - K\epsilon}{K}\right) \cdot \left(\frac{c}{\|P\|} - \epsilon\right) > K\|P\| \ .$$

Now in any n-dimensional Banach space, for any $\epsilon > 0$, $\left(\frac{8}{\epsilon}\right)^n$ is an upper bound on the cardinality of a minimal $\frac{\epsilon}{4}$-net over the unit ball. For such an $\frac{\epsilon}{4}$-net, the family of balls with centers in the net and radii $\frac{\epsilon}{2}$ is a covering of the unit ball. It follows that any subset A of the unit ball with $\bar{\bar{A}} \geq m\left(\frac{8}{\epsilon}\right)^n$ elements contains m vectors $y_1, y_2, \cdots, y_m$ such that $\|y_i - y_l\| < \epsilon, \forall i, l = 1, 2, \cdots, m$. Now choose $j \in \mathbb{N}$ so that

$$\exp_j(n) \geq m\left(\frac{8}{\epsilon}\right)^n,$$

and note that j is independent of n, but depends upon $K, \|P\|$, and c. Assume that the lemma fails for this j and some Banach space X of dimension n. By re-indexing the $\{z_i\}_{i=1}^{k}$ (if necessary), we may assume that $\exists$ an

$$s > \exp_j(n) \geq m\left(\frac{8}{\epsilon}\right)^n \text{ so that}$$

60

$$|z_i^*(x_i)| \geq c, \quad \text{for} \quad i = 1, 2, \cdots, s \ .$$

It follows (perhaps after another re-indexing) that $\exists z_1, z_2, \cdots, z_m$ so that

a). $|z_i^*(x_i)| \geq c$, and

b). $\|x_i - x_l\| < \epsilon, \forall i, l = 1, 2, \cdots, m.$

Now consider the projection P_0 on Z defined by:

$$P_0(z) := \sum_{i=1}^{m} z_i^*(z) z_i, \quad (z \in Z).$$

By the definition of P_0 and the fact that $\{z_i\}_{i=1}^{k}$ is K-unconditional, we have:

$$\|P_0 x_1\| \leq \|P\| \cdot K \|x_1\| \leq K \|P\| \ . \tag{1}$$

Also for $1 \leq i \leq m$, letting x_i^* denote the coefficient functional associated with x_i, we have:

$$|x_i^*(x_1)| \geq |x_i^*(x_i)| - |x_i^*(x_i - x_1)| \geq |z_i^*(x_i)| - K\epsilon \geq c - K\epsilon,$$

since the K-unconditionality of $\{z_i\}_{i=1}^{k}$ implies that $\|z_i^*\| \leq K$, and $\|x_i^*\| \leq K, \quad (i = 1, 2, \cdots, k)$. Applying this inequality, we have

$$
\begin{aligned}
\|P_0 x_1\| &= \left\| \sum_{i=1}^{m} z_i^*(x_1) z_i \right\| \\
&= \left\| \sum_{i=1}^{m} x_i^*(x_1) z_i \right\| \\
&\geq \frac{c - K\epsilon}{K} \left\| \sum_{i=1}^{m} z_i \right\| \\
&\geq \frac{c - K\epsilon}{K} \left\| \sum_{i=1}^{m} x_i \right\| \\
&\geq \frac{c - K\epsilon}{K} \left[m\|x_1\| - \left\| \sum_{i=1}^{m} (x_1 - x_i) \right\| \right] \\
&\geq \frac{c - K\epsilon}{K} \left[\frac{mc}{\|P\|} - m\epsilon \right] \\
&= m \left(\frac{c - K\epsilon}{K} \right) (c\|P\| - \epsilon) > K\|P\| \ .
\end{aligned}
$$

But this would contradict inequality (1). It follows that

$$s \leq \exp_j(n), \text{ and that}$$

$$\bar{\bar{A}} \leq \exp_j(n) = \exp_j(\dim X) \ .$$

$\square$

Here (and in later chapters) we need a result of J. Bourgain, P. Casazza, J. Lindenstrauss, and L. Tzafriri [15]. We state it here (for the sake of completeness), but omit the proof.

Lemma VII.a.2: Let X, Y be Banach spaces, P a projection of $X \oplus Y$ onto a subspace Z which has a K-unconditional basis $\{z_i\}_{i=1}^{n}$. Let $\{z_i^*\}_{i=1}^{n} \subset Z^*$ be such that

$$Pz = \sum_{i=1}^{n} z_i^*(z) z_i, \quad (z \in X \oplus Y) \ .$$

Also let $E := \left\{1 \leq i \leq n : |z_i^*(x_i)| \geq \frac{1}{2}\right\}$, (where each z_i is decomposed as $z_i = x_i + y_i$, with $x_i \in X, y_i \in Y$). Then there is a function $M\,(:= M\,(K, \|P\|))$ such that $[z_i : i \in E]$ is M-equivalent to an M-complemented subspace of $(\sum \oplus X)_{l_p}$, (where there are $2^{\bar{E}}$ summands) and this is true for all $1 \leq p < \infty$ (and in fact, with c_0 in lieu of l_p).

Our next result is a generalization of Proposition V.13, and although it appears rather formidable, its proof is easily motivated and its content will be of fundamental importance to our study in Chapter IX of complemented subspaces of T which have unconditional bases.

Lemma VII.a.3: There is a function $f : [1, \infty) \times [1, \infty) \rightarrow [1, \infty)$ such that for each n, whenever P is a projection from $Z := (X_n \oplus X_n \oplus \cdots)_{l_1}$, where $X_n := [t_i]_{i=n}^{\infty}$, onto an n-dimensional subspace with a normalized K-unconditional basis $\{x_i\}_{i=1}^n$, then $\{x_i\}_{i=1}^n$ is $f\,(K, \|P\|)$-equivalent to the unit vector basis of l_1^n.

Proof: By Proposition V.6, there exist

a). an $\epsilon > 0$, (depending upon $\|P\|$),

b). a natural number $s_n := \left[\left[\frac{2n^2}{\epsilon}\right]\right]^n$, (where $[[\cdot]]$ is the greatest integer function),

and

c). s_n normalized disjointly supported vectors $\{y_i : 1 \leq i \leq s_n\} \subset Z$

such that PZ is 2-isomorphic to a $2\|P\|$-complemented subspace of $[y_i : 1 \leq i \leq s_n]$.

It follows from Proposition V.11 that any such vectors y_i are M-equivalent to the unit vector basis of $l_1^{s_n}$ (where M depends upon the above ϵ). It easily follows that this also holds in Z, i.e. $\exists$ a function $f_0(\|P\|)$ such that $\{y_i : 1 \leq i \leq s_n\}$ is $f_0(\|P\|)$-equivalent to the unit vector basis of $l_1^{s_n}$.

Since PZ is 2-isomorphic to a $2\|P\|$-complemented subspace of $[y_i : 1 \leq i \leq s_n]$, it follows that $\exists$ a function $f_1(\|P\|)$ such that PZ is $f_1(\|P\|)$-isomorphic to a $f_1(\|P\|)$-complemented subspace of l_1. Since $\{x_i\}_{i=1}^n$ is a K-unconditional basis for this complemented subspace PZ of l_1, it follows that $\exists$ a function $f(K, \|P\|)$ such that $\{x_i\}_{i=1}^n$ is $f(K, \|P\|)$-equivalent to the unit vector basis of l_1^n. $\square$

We can now state and prove the main tool for studying bounded operators on subsequences of $\{t_n\}_{n=1}^{\infty}$.

Proposition VII.a.4: There is a function $f : [1, \infty) \times [1, \infty) \rightarrow [1, \infty)$ with $\lim_{x \to \infty} f(x, y) = \infty$ for each fixed $y > 1$ which satisfies the following:

for any natural numbers $n < m$, if we let $W := [t_i]_{i=n}^m$, if we let $L : W \rightarrow [t_i : 1 \leq i < n, \text{ or } i > n]$ be an invertible operator, and if P is a projection of T onto $L(W)$, then

$$\|L\| \cdot \|L^{-1}\| \geq f\,(\|I\|, \|P\|)$$

where I is the formal identity from W into l_1.

Proof: Fix $n < m$, let W be as above, and let $X := [t_i]_{i=1}^{n-1}$ and $Y := [t_i]_{i=m+1}^{\infty}$. Assume that L is an invertible operator $L : W \rightarrow X \oplus Y$, and that P is a projection of T onto $L(W)$. For each $n \leq i \leq m$, let $Lt_i = x_i + y_i$, where $x_i \in X, y_i \in Y$. Now choose coefficient functionals $(Lt_n)^*, (Lt_{n+1})^*, \cdots, (Lt_m)^*$, so that

$$Pz = \sum_{i=n}^{m} (Lt_i)^*(z) \cdot Lt_i, \qquad (z \in T) .$$

Define

$$E := \left\{ n \leq i \leq m : |(Lt_i)^*(y_i)| \geq \tfrac{1}{2} \right\}, \quad \text{and}$$
$$F := \{ n \leq i \leq m : i \notin E \} .$$

By Lemma VII.a.2, $\exists$ a function $M := M(K, \|P\|)$ such that $[Lt_i : i \in E]$ is M-equivalent to an M-complemented subspace of $(\sum \oplus Y)_{l_1}$, where we have $2^{\bar{\bar{E}}}$ summands (and where K is the unconditionality constant of $\{Lt_i\}_{i=n}^{m}$).

By Lemma VII.a.3, $\exists$ a function $f_0 : [1, \infty) \times [1, \infty) \rightarrow [1, \infty)$ such that $\{Lt_i : i \in E\}$ is $f_0(K, \|P\|)$-equivalent to the unit vector basis of $l_1^{\bar{\bar{E}}}$. Since $F \subset \left\{ n \leq i \leq m : |(Lt_i)^*(x_i)| \geq \tfrac{1}{2} \right\}$, Lemma VII.a.1 implies $\exists$ a natural number j (depending on $K, \|P\|$, where $c = \tfrac{1}{2}$ here) so that

$$\bar{\bar{F}} \leq \exp_j(n-1) .$$

By Proposition IV.b.1, $\{t_i : i \in F\}$ is $\|L\| \cdot \|L^{-1}\| \cdot 2^{j+1}$-equivalent to the unit vector basis of $l_j^{\bar{\bar{F}}}$.

Combining these facts, we have that $\{Lt_i\}_{i=n}^{m}$ is $2^{j+1} \cdot \|L\| \cdot \|L^{-1}\| \cdot f_0(K, \|P\|)$-equivalent to the unit vector basis of l_1^{m-n+1}. Since K is a function of $\|L\| \cdot \|L^{-1}\|$, and $\|I\|$ is the equivalence constant between $\{t_i\}_{i=n}^{m}$ and the unit vector basis of l_1^{m-n+1}, it follows that $\exists$ a function $f(\|I\|, \|P\|)$ such that $\|L\| \cdot \|L^{-1}\| \geq f(\|I\|, \|P\|)$. (It is clear that $\lim_{x \to \infty} f(x, y) = +\infty$, for each fixed $y > 1$.) $\square$

Notes and Remarks:

1/. Proposition VII.a.4 should be compared to Proposition VI.b.2. The former is a strengthening of the latter, obtained by assuming that the operator L has complemented range.

2/. The results used to prove Proposition VII.a.4 also hold for subsequences $\{t_{k_n}\}_{n=1}^{\infty}$ of $\{t_n\}_{n=1}^{\infty}$, so we also have a valid variant of Proposition VII.a.4 when we replace W by $[t_{k_i}]_{i=n}^{m}$.

3/. Note that in Lemma VII.a.1, for $\|P\|$ amd c fixed, j increases with K.

B. Isomorphic subspaces of T spanned by subsequences of $\{t_n\}_{n=1}^{\infty}$.

We are now in a position to address the question of when two subsequences of $\{t_n\}_{n=1}^{\infty}$ span isomorphic subspaces. Recall from the theory of Schauder bases:

Definition VII.b.1: If $\{x_n\}_{n=1}^{\infty}$ and $\{y_n\}_{n=1}^{\infty}$ are bases, then $\{x_n\}_{n=1}^{\infty}$ <u>dominates</u> $\{y_n\}_{n=1}^{\infty}$, if for all choices of scalars $\{a_n\}_{n=1}^{\infty}$, whenever $\sum_n a_n x_n \in [x_n]_{n=1}^{\infty}$, then $\sum_n a_n y_n \in [y_n]_{n=1}^{\infty}$. We write:

$$\text{``}\{x_n\}_{n=1}^{\infty} >> \{y_n\}_{n=1}^{\infty}\text{''} .$$

With this notion, and tools from the last section, we obtain:

Theorem VII.b.2: Let $\{t_{k_n}\}_{n=1}^{\infty}$ and $\{t_{j_n}\}_{n=1}^{\infty}$ be subsequences of $\{t_n\}_{n=1}^{\infty}$ such that $[t_{k_n}]_{n=1}^{\infty}$ is isomorphic to a complemented subspace of $[t_{j_n}]_{n=1}^{\infty}$.

$$\text{Then} \quad \{t_{k_n}\}_{n=1}^{\infty} >> \{t_{j_n}\}_{n=1}^{\infty} \ .$$

Proof: Let L be an isomorphism sending $[t_{k_n}]_{n=1}^{\infty}$ into $[t_{j_n}]_{n=1}^{\infty}$, and let P be a projection of $[t_{j_n}]_{n=1}^{\infty}$ onto the range of L. (Without loss of generality, we may assume that P is a projection on T.)

Let $E_i := \{k_n : j_{i-1} < k_n \leq j_i\}$, (where we let $j_0 := 0$), for $i = 1, 2, \cdots$. Let I_i be the natural identity map from $[t_m : m \in E_i]$ into l_1. Let $f(x, y)$ be the function given by Remark 2 (of "Notes and Remarks" of Chapter VII.A) for the subsequence $\{t_{k_n}\}_{n=1}^{\infty}$ of $\{t_n\}_{n=1}^{\infty}$. Upon restricting L to $[t_m : m \in E_i]$, we see that $L([t_m : m \in E_i])$ is $\|L\| \cdot \|L^{-1}\|$-isomorphic to a $\|L\| \cdot \|L^{-1}\| \cdot \|P\|$-complemented subspace of $[t_{j_n}]_{n=1}^{\infty}$. Therefore Proposition VII.a.4 yields an appropriate f such that:

$$f\left(\|I_i\|, \|L\| \cdot \|L^{-1}\| \cdot \|P\|\right) \leq \|L\| \cdot \|L^{-1}\|, \quad (i = 1, 2, \cdots) \ .$$

Since $\lim_{x \to \infty} f(x, y) = \infty$, for each fixed $y > 1$, it follows that $\sup_i \|I_i\| < \infty$. By the proof of Corollary IV.a.2, it follows that $\{t_{k_n}\}_{n=1}^{\infty}$ is equivalent to a subsequence of $\{t_{j_n}\}_{n=1}^{\infty}$, and hence that $\{t_{k_n}\}_{n=1}^{\infty} >> \{t_{j_n}\}_{n=1}^{\infty}$. $\square$

Corollary VII.b.3: Let $\{t_{k_n}\}_{n=1}^{\infty}$ and $\{t_{j_n}\}_{n=1}^{\infty}$ be subsequences of $\{t_n\}_{n=1}^{\infty}$. Then the spaces $[t_{k_n}]_{n=1}^{\infty}$ and $[t_{j_n}]_{n=1}^{\infty}$ are isomorphic iff $\{t_{k_n}\}_{n=1}^{\infty}$ is equivalent to $\{t_{j_n}\}_{n=1}^{\infty}$.

Proof: Two basic sequences are equivalent iff they dominate one another. $\square$

Our next application of Proposition VII.a.4 shows that a (bounded) percentage of every unconditional basis for an n-dimensional complemented subspace of T is equivalent to a subset of the unit vector basis of l_1^n. This generalizes the obvious fact that for any sequence $\{y_i\}_{i=1}^n$ of normalized disjointly supported vectors in T, $\exists E \subset \{1, 2, \cdots, n\}$ such that $\bar{\bar{E}} \geq \frac{n}{2}$ and $\{y_i : i \in E\}$ is 2-equivalent to the unit vector basis of $l_1^{\bar{\bar{E}}}$.

Theorem VII.b.4: There is a function $h : (0,1) \times [1, \infty) \times [1, \infty) \to [1, \infty)$ satisfying the following:

for every $0 < c < 1$ and any normalized K-unconditional basis $\{x_i\}_{i=1}^n$ for an n-dimensional $\|P\|$-complemented subspace of T, $\exists E \subset \{1, 2, \cdots, n\}$ such that

i) $\bar{\bar{E}} \geq cn$, and

ii) $\{x_i : i \in E\}$ is $h(c, K, \|P\|)$-equivalent to the unit vector basis of $l_1^{\bar{\bar{E}}}$.

Proof: Let P be a projection of T onto an n-dimensional subspace X with normalized K-unconditional basis $\{x_i\}_{i=1}^n$. Choose j (depending on K, $\|P\|$, and c) independent of n, as given by Lemma VII.a.1. Now choose k beyond j so that: $\left[\frac{c}{1-c} \exp_j(m+1)\right] + 1 \leq \exp_k(m), (m =$

$1, 2, \cdots, $). Since k depends only upon c, by Corollary IV.b.2, there is a function $f(c)$ such that if L is the operator $L : T \to [t_{\exp_k(m)}]_{m=1}^\infty$, given by $Lt_m = t_{\exp_k(m)}$, (and extended by linearity), then $\|L\| \cdot \|L^{-1}\| = f(c)$.

For the n above, choose $m \in \mathbb{N}$ so that:

$$\exp_j(m) \le [[(1-c)n]] \le (1-c)n < \exp_j(m+1).$$

Now let

$$Z := [t_{\exp_k(i)}]_{i=1}^m ,$$

and

$$Y := [t_{\exp_k(i)}]_{i=m+1}^\infty .$$

Applying the isomorphism L to $\{x_i\}_{i=1}^n$, without loss of generality we may assume that:

a). $x_i \in Z \oplus Y, \quad (i = 1, 2, \cdots, n)$,

b). $[x_i]_{i=1}^n$ is $\|L\| \cdot \|L^{-1}\| \cdot \|P\| =: \|Q\|$-complemented in $Z \oplus Y$,(where Q is the obvious composition), and

c). $\{x_i\}_{i=1}^n$ is a normalized $\|L\| \cdot \|L^{-1}\| \cdot K$-unconditional basis for X.

Decompose each x_i as $x_i = z_i + y_i$ against Z and Y, and choose $x_i^* \in (Z \oplus Y)^*, \quad (i = 1, 2, \cdots, n)$, such that

$$Qx = \sum_{i=1}^n x_i^*(x)x_i, \quad (x \in Z \oplus Y) .$$

Define now

$$
\begin{aligned}
H &:= \left\{ 1 \le i \le n : |x_i^*(y_i)| \ge \tfrac{1}{2} \right\}, \text{ and} \\
F &:= \{ 1 \le i \le n : i \notin H \} .
\end{aligned}
$$

Since $\dim Z = m$, Lemma VII.a.1 yields that

$$\bar{\bar{F}} \le \exp_j(m) \le [[(1-c)n]] .$$

Therefore $\bar{\bar{H}} \ge cn$, and we can choose a subset $E \subset H$ such that $cn \le \bar{\bar{E}} \le cn + 1$. Since $(1-c)n \le \exp_j(m+1)$, it follows that

$$\bar{\bar{E}} \le cn + 1 \le \left[\frac{c}{1-c} \cdot \exp_j(m+1) \right] + 1 \le \exp_k(m) .$$

By Lemma VII.a.2, $[x_i : i \in E]$ is M-equivalent to an M-complemented subspace of $(\sum \oplus Y)_{l_1}$, (with $2^{\bar{\bar{E}}}$ summands here), where M depends on $\|L\|, \|L^{-1}\|, K$, and $\|P\|$. Since $f(c) = \|L\| \cdot \|L^{-1}\|$, Lemma VII.a.3 now yields the existence of a function h so that $\{x_i : i \in E\}$ is $h(c, K, \|P\|)$-equivalent to the unit vector basis of $l_1^{\bar{\bar{E}}}$. $\square$

Notes and Remarks:

1/. All of the results of this chapter dualize to results about T^*. Upon replacing l_1 by c_0, and l_1^n by l_∞^n throughout, the proofs presented suffice.

2/. It is easily shown that Lemma VII.a.1 cannot be strengthened. Indeed, with the notation of that Lemma, if $i \in A$, then the condition on A implies that $\|x_i\| \geq c$. However, if we set $B := \{i : \|x_i\| \geq c\}$, there need not exist a j (depending on $K, \|P\|$, and c) such that for all Banach spaces X,

$$\bar{\bar{B}} \leq \exp_j(\dim X) \ .$$

3/. Since T^* is minimal, Theorem VII.b.2 would fail in T^* without the (statement analogous to the) hypothesis that $[t_{k_n}]_{n=1}^\infty$ be complemented in $[t_{j_n}]_{n=1}^\infty$. (It may be that this hypothesis is unnecessary in T, but we do not know how to prove this conjecture.)

4/. Theorem VII.b.4 fails without the assumption of complementation. This is a consequence of Dvoretsky's Theorem on Spherical Sections: For each infinite-dimensional normed linear space X, each $n \geq 1$, and each $\epsilon > 0, \exists$ a one-one linear operator L from l_2^n into X such that $\|L\| \cdot \|L^{-1}\| < 1 + \epsilon$.

5/. Recall:

Definition VII.b.5: A Banach space X is <u>sufficiently Euclidean</u> if it contains:

1). a sequence of subspaces $\{E_n\}_{n=1}^\infty$ for which $\sup_n d(E_n, l_2^n) < \infty$, and

2). $\exists$ a sequence of projections $\{P_n\}_{n=1}^\infty$ from X onto E_n such that $\sup_n \|P_n\| < \infty$.

C. Stegall and J. Rutherford [47] asked whether every reflexive Banach space is sufficiently Euclidean. The following is an immediate corollary of Theorem VII.b.4:

Theorem VII.b.6: <u>Tsirelson's space T is not sufficiently Euclidean.</u> (This was first proved by W. Johnson [27].)

Chapter VIII: Permutations of the unit vector basis of Tsirelson's space.

In this chapter we will study the effects that permutations of the unit vector basis have upon the norm in Tsirelson's space.

A. The non-increasing rearrangement operator D.

Definition VIII.a.1: For $x = \sum_n a_n t_n \in T$, the non-increasing rearrangement of x is the vector $\hat{x} := \sum_n b_n t_n$, where $\{b_n\}_{n=1}^\infty$ is the non-increasing rearrangement of the non-zero elements of $\{|a_n|\}_{n=1}^\infty$. The non-increasing rearrangement operator is the operator $D : T \to T$ defined by $Dx = \hat{x}, (x \in T)$.

D is clearly a non-linear operator, and $\|D\| > 1$, as can be seen by calculating some examples:

Let $x = (7, 7, 7, 4, 7, 4, 4, 0, 0, \cdots) \in T$. Then $\hat{x} = (7, 7, 7, 7, 4, 4, 4, 0, 0, \cdots)$, and some (tedious) calculations yield that $\|x\| = \frac{19}{2}$, while $\|\hat{x}\| = 10$. We will show that D is a bounded operator on T, and will simplify the argument by first proving several lemmas concerning special types of rearrangements of coefficients and basis vectors in T.

Our first lemma is a generalization of Proposition I.9(3).

Lemma VIII.a.2: Let $\{t_n\}_{n=1}^\infty$ be the unit vector basis of T_M (the "modified" Tsirelson's space of Chapter V), and let $n_1 < n_2 < \cdots$, and $\{m_i\}_{i=1}^\infty$ be natural numbers such that $m_i \leq n_i, \forall i$. Define a linear operator

$$L : [t_{n_i}]_{i=1}^\infty \to T_M \text{ by } Lt_{n_i} := t_{m_i}, \forall i \quad .$$

Then:

$$\|Lx\|_M \leq \|x\|_M, \forall x \in T_M \quad .$$

Proof: We will show that $\|Lx\|_{M,m} \leq \|x\|_M$, for $m = 0, 1, 2, \cdots$. The case $m = 0$ is obvious, so we assume this inequality for some fixed m and for all $x \in [t_{n_i}]_{i=1}^\infty$. Choose $x = \sum_i a_i t_{n_i} \in [t_{n_i}]_{i=1}^\infty, k \geq 1$, and disjoint sets $\{E_j\}_{j=1}^k$ in $\mathbb{N}$ with $k \leq E_j, (j = 1, 2, \cdots, k)$. By the induction hypothesis applied to each summand in the second sum below, we obtain

$$\frac{1}{2} \sum_{j=1}^k \|E_j Lx\|_{M,m} = \frac{1}{2} \sum_{j=1}^k \|\sum \{a_i t_{m_i} : m_i \in E_j\}\|_{M,m} \tag{$(*)$}$$

$$\leq \frac{1}{2} \sum_{j=1}^k \|\sum \{a_i t_{n_i} : m_i \in E_j\}\|_M \quad .$$

Our assumption that $m_i \leq n_i, \forall i$, implies that $k \leq n_i$ whenever $m_i \in E_j$, for some $j = 1, 2, \cdots, k$. Thus the last sum in (*) is allowable for x.

Thus

$$\frac{1}{2} \sum_{j=1}^k \|E_j x\|_{M,m} \leq \|x\|_M.$$

It now follows that $\|Lx\|_{M,m+1} \leq \|x\|_M$, and the induction is complete. $\square$

Lemma VIII.a.2 implies that any operator from $[t_{n_i}]_{i=1}^{\infty}$ to T_M which moves basis vectors "to the left" has norm 1.

Remark VIII.a.3: When applying the above lemma, we will find it convenient to think of the operator L as being induced by the mapping $\sigma : \{n_i : i \in \mathbb{N}\} \to \mathbb{N}$ given by $\sigma(n_i) = m_i, \forall i$. In this setting, we can rephrase Lemma VIII.a.2 as:

If $A = \{n_1 < n_2 < \cdots\}$ and $\sigma : A \to \mathbb{N}$ is a one-one map for which $\sigma(n_i) \le n_i, \forall i$, then:

$$\| \sum_i a_i t_{\sigma(n_i)} \|_M \le \| \sum_i a_i t_{n_i} \|_M,$$

for all sequences of scalars $\{a_i\}_{i=1}^{\infty}$.

We will also need the following facts:

Lemma VIII.a.4: If $3 \le k_1 < k_2 < \cdots < k_s$, $A = \left\{ 2^{k_1}, 2^{k_2}, \cdots, 2^{k_s} \right\}$, and if $B = \{p+1, p+2, \cdots, p+s\}$, for some p and s, and if $\sigma : A \to B$ satisfies $\sigma(i) \ge i, (i \in A)$, then $3s \le p$.

Proof: By hypothesis, $2^{k_s} \le \sigma(2^{k_s}) \le p + s$. Thus $p \ge 2^{k_s} - s \ge 2^{s+2} - s \ge 3s$. $\square$

Lemma VIII.a.5: Let $F_1 < E_1 < F_2 < E_2 < \cdots < F_s < E_s$ be subsets of natural numbers with $p_i := \bar{\bar{E}}_i$, and $q_i := \bar{\bar{F}}_i$. Assume that $\sum_{i=1}^{s} q_i \ge \sum_{i=1}^{s} p_i$, and that

$$\sum_{i=j}^{s} q_i < \sum_{i=j}^{s} p_i, \quad (j = 2, 3, \cdots, s) .$$

Let

$$A := \bigcup_{i=1}^{s} E_i = \{m_1 < m_2 < \cdots < m_k\} ,$$

and

$$\bigcup_{i=1}^{s} F_i = \{n_1 < n_2 < \cdots < n_r\} .$$

Then clearly $k \le r$, and so $B := \{n_{r-k+1} < n_{r-k+2} < \cdots < n_r\}$ has the property $\bar{\bar{A}} = \bar{\bar{B}}$. Also:

$$n_{r-k+i} \le m_i, \quad (i = 1, 2, \cdots, k) .$$

Proof: (We prove only the final claim.)

We count backwards from r : since $F_s < E_s$, $n_{r-q_s+1} < n_{r-q_s+2} < \cdots < n_r < m_{k-q_s+1} < m_{k-q_s+2} < \cdots < m_k$.

Now since $q_s < p_s$,

$$n_{r-p_s+i} \in F_{s-1}, \text{ and } m_{r-p_s+i} \in E_s ,$$
$$(\text{both for } i = 1, 2, \cdots, p_s - q_s) .$$

Also, $F_{s-1} < E_{s-1} < E_s$ implies that:

$$n_{r-p_s+1} < n_{r-p_s+2} < \cdots < n_{r-q_s} < m_{k-p_s+1} < m_{k-p_s+2} < \cdots < m_{k-q_s} .$$

Now continue "counting backwards" to see that $n_{r-k+i} \le m_i$, $(i = 1, 2, \cdots, k)$. $\square$

Our next lemma is quite technical, so a word about our strategy is in order. Our main goal in this section is to prove:

Claim a: D is a bounded operator.

We maintain that in order to demonstrate this, it suffices to show:

Claim b: $\exists c > 0$ such that for each $x = \sum_n a_n t_n \in T$ where $a_n > 0$, $(n = 1, 2, \cdots), \|Dx\| \leq c\|x\|$.

Note that if Claim b holds, and if $x = \sum_n a_n t_n \in T$, choose $m_1 < m_2 < \cdots$ such that $\{m_1, m_2, \cdots\} = \operatorname{supp}(x)$. Then letting $y = \sum_i |a_{m_i}| t_i$, note that

$$\|Lx\| = \|Ly\| \leq c\|y\| \leq c\|x\|, \tag{**}$$

where L is the operator of Remark VIII.a.3. (The equality in (**) holds since the moduli of the coefficients of Lx and Ly are exactly the same. The first inequality holds by our assumption on L, while the last comes from Proposition I.9(3).) It follows that for any $x = \sum_n a_n t_n \in T$, with all $a_n > 0$, that $\exists$ a permutation σ of $\mathbb{N}$ so that

$$Lx = \sum_n a_{\sigma^{-1}(n)} t_n .$$

By Lemma VIII.a.2, if $\sigma^{-1}(n) \leq n, \forall n$, then we're done, since $\|Lx\| \leq \|x\|$. This is, of course, not necessarily true in general, so we set:

$$
\begin{aligned}
x_1 &:= \sum \left\{ a_{\sigma^{-1}(n)} t_n : \sigma^{-1}(n) \leq n \right\}, \text{and} \\
x_2 &:= \sum \left\{ a_{\sigma^{-1}(n)} t_n : \sigma^{-1}(n) > n \right\},
\end{aligned}
$$

and note that $Lx = x_1 + x_2$, and that (by Lemma VIII.a.2), $\|x_1\| \leq \|x\|$. So if we can produce a universal constant $c_1 > 0$ such that

$$\|x_2\| \leq c_1 \|x_1\|, \forall x \in T,$$

then we could conclude: $\|Lx\| \leq (1 + c_1)\|x\|$, for all $x \in T$.

In other words, we would have shown that the coefficients which L "moves to the right" have smaller norm (up to a universal constant) than the coefficients which L "moves to the left".

To make it easier to show this fact, we introduce an intermediate transformation L_1 which maps t_n to t_{2^n}, for all n. By (3) of the "Notes and Remarks" of Chapter III, L_1 is an isomorphism on T. Since

$$
\begin{aligned}
\operatorname{supp}(L_1 x) &= \{2^n : n \in \mathbb{N}\}, \text{and} \\
\operatorname{supp}(DL_1 x) &= \mathbb{N},
\end{aligned}
$$

D will map "most" of the coefficients of $L_1 x$ to the left, and thus (as we will show) the coefficients it maps "right-ward" will contribute little to the norm of Lx. It will then follow that D is bounded on $L_1 T$ by a constant $c > 0$, and so finally:

$$\|Dx\| = \|DL_1 x\| \leq c\|L_1 x\| \leq c\|L_1\|\|x\|, \quad \forall x \in T.$$

To show that $D|_{L_1T}$ maps "most" elements left-wards, we will choose sets $F_1 < E_1 < F_2 < E_2 \cdots$ so that

$$Dx = \sum_i (E_i x + F_i x), \quad \text{and} \quad \forall n,$$

$$\sigma^{-1}(n) \in E_j, \text{ (for any } j \text{), implies } \sigma^{-1}(n) \leq n, \quad \text{and}$$

$$\sigma^{-1}(n) \in F_j, \text{ (for any } j \text{), implies } \sigma^{-1}(n) > n.$$

Then we will use a generalization of the counting argument of Lemma VIII.a.4 to compare $\bar{\bar{E}}_i$ to $\bar{\bar{F}}_i$. Our next lemma provides the required generalization.

Lemma VII.a.6: Let s, m, and $n \in \mathbb{N}$ and assume that

$$\{m, m+1, \cdots, m+n\} = (\cup_{i=1}^s E_i) \cup (\cup_{i=1}^s F_i),$$

where the E_i and F_i satisfy the following:

i). $\bar{\bar{E}}_i =: p_i$, and $\bar{\bar{F}}_i =: q_i$, $(i = 1, 2, \cdots, s)$,

ii). $F_1 < E_1 < F_2 < E_2 < \cdots < F_s < E_s$.

iii). $q_1 \leq \sum_{i=1}^s p_i$,

iv). $\sum_{i=j}^s q_i \leq \sum_{i=j}^s p_i$, $(j = 2, 3, \cdots, s)$,

v). $\sum_{i=1}^s q_i \geq \sum_{i=1}^s p_i$.

Let $l := \sum_{i=1}^s p_i$, $3 < k_1 < k_2 < \cdots$ be natural numbers, and σ a bijection from $A := \left\{ 2^{k_1}, 2^{k_2}, \cdots, 2^{k_l} \right\}$ onto $\cup_{i=1}^s E_i$ such that $\sigma(i) \geq i$, for all $i \in A$.

Then: $\min F_1 \geq 2l$, and we have the following inequalities for all sequences of scalars $\{a_i\}_{i=1}^\infty$:

a). $\left\| \sum_{j=1}^s \sum \{a_i t_i : i \in E_j\} \right\| \geq \frac{1}{2} \sum_{j=1}^s \sum \{|a_i| : i \in E_j\}$,

b). $\left\| \sum_{j=1}^s \sum \{a_i t_i : i \in F_j\} \right\| \geq \frac{1}{4} \sum_{j=1}^s \sum \{|a_i| : i \in F_j\}$,

Proof: Since $m + n = \max$ range σ, and $\sigma(i) \geq i$ over A, it follows that

$$m + n \geq \sigma(2^{k_l}) \geq 2^{k_l} \geq 2^{l+3} .$$

(We've also used here the assumption that $3 < k_1 < k_2 < \cdots$).

Also, (iii) and (iv) imply that

$$l + \sum_{i=1}^s q_i = l + q_1 + \sum_{i=2}^s q_i \leq l + l + l = 3l.$$

Combining these results, we get

$$\min F_1 = m + n - \sum_{i=1}^s p_i - \sum_{i=1}^s q_i \geq 2^{l+3} - 3l \geq 2l . \tag{1}$$

To show (a), note that

$$2l = 2\sum_{i=1}^{s} p_i \le \min F_1 < E_1 < E_2 < \cdots < E_s, \text{ and } \overline{\cup_{i=1}^{s} E_i} = l .$$

Hence, $\| \sum_{j=1}^{\infty} \sum \{a_i t_i : i \in E_j\} \| \ge \frac{1}{2} \sum_{j=1}^{s} \sum \{|a_i| : i \in E_j\}$, since the right side of this inequality is

admissible for the vector in the left. To see (b) note that (by Lemma VIII.a.4) $\frac{1}{2} \sum \{|a_i| : i \in F_1\}$

is admissible for $\sum \{a_i t_i : i \in F_1\}$. Also (by assumption (iv) and (1) above), we have:

$$\overline{\cup_{i=2}^{s} F_i} = \sum_{i=2}^{s} \overline{F_i} \le \sum_{i=2}^{s} p_i \le l \le \min F_1 < \min F_2 .$$

Therefore, $\frac{1}{2} \sum_{j=2}^{s} \sum \{|a_i| : i \in F_j\}$ is an admissible sum for

$$\sum_{j=2}^{s} \sum \{a_i t_i : i \in F_j\} \in T .$$

It follows that:

$$\| \sum_{j=1}^{s} \sum \{a_i t_i : i \in F_j\} \|$$
$$\ge \frac{1}{2} \left(\| \sum \{a_i t_i : i \in F_1\} \| + \sum_{j=2}^{s} \| \sum \{a_i t_i : i \in F_j\} \| \right)$$
$$\ge \frac{1}{4} \sum_{j=1}^{s} \sum \{|a_i| : i \in E_j\} .$$

$\square$

Our next lemma will show that the norm of that portion of Dx whose coefficients were "moved to the right" is smaller than a universal constant times the norm of that portion of Dx whose coefficients were "moved to the left", if $x \in [t_{2^i}]_{i=1}^{\infty}$. (This is the last result needed for our proof that D is a bounded operator on T).

Lemma VIII.a.7: For any $x = \sum_{n=4}^{\infty} a_{2^n} t_{2^n} \in T$ with $a_{2^n} > 0$, $(n = 4, 5, \cdots)$, if $Dx := \sum_n b_n t_n$ is the non-increasing rearrangement of x and σ is a bijection from $\{2^4, 2^5, \cdots\}$ onto $\mathbb{N}$ such that $b_n = a_{\sigma^{-1}(n)}, \forall n$, then

$$\| \sum \{b_n t_n : \sigma^{-1}(n) < n\} \| \le 2^4 \cdot 3^5 \| \sum \{b_n t_n : \sigma^{-1}(n) \ge n\} \| .$$

Proof: It suffices to prove the lemma for finitely supported vectors.

So assume

$$x = \sum_{n=4}^{m+3} a_{2^n} t_{2^n},$$
$$y := Dx = \sum_{i=1}^{m} b_i t_i,$$
$$\sigma : \{2^4, 2^5, \cdots, 2^{m+3}\} \to \{1, 2, \cdots, m\} \text{ a bijection,}$$
$$\text{and } b_n = a_{\sigma^{-1}(n)}, \quad (n = 1, 2, \cdots, m).$$

Note that $\sigma^{-1}(1) \ge 1$, and choose n_1 as the smallest natural number such that

$$\sigma^{-1}(n_1) < n_1 .$$

Let
$$F_1 : \{1, 2, \cdots, n_1 - 1\} \ .$$

Now choose n_2 as the smallest natural number beyond n_1 for which $\sigma^{-1}(n_2) \geq n_2$, and let
$$E_1 := \{n_1, n_1 + 1, \cdots, n_2 - 1\} \ ,$$

and iterate this process to produce sets
$$1 \leq F_1 < E_1 < F_2 < E_2 < \cdots < F_s < E_s$$

with the properties
$$\left(\cup_{j=1}^s F_j\right) \cup \left(\cup_{j=1}^s E_j\right) = \{1, 2, \cdots, m\}, \tag{2}$$

$$\text{if } i \in F_j, \text{ (for some } j = 1, 2, \cdots, s), \text{ then } \sigma^{-1}(i) \geq i, \text{ and} \tag{3}$$

$$\text{if } i \in E_j, \text{ (for some } j = 1, 2, \cdots, s), \text{ then } \sigma^{-1}(i) < i \ . \tag{4}$$

It follows that $1 + \max F_j = \min E_j$, $(j = 1, 2, \cdots, s)$, and that each of these sets is an interval in $\mathbb{N}$.

Now let $\overline{\overline{E_j}} =: p_j$ and $\overline{\overline{F_j}} =: q_j$, $(j = 1, 2, \cdots, s)$. Next we divide these sets into groups, each of which satisfies Lemma VIII.a.6.

To do this, first observe that if $1 \leq l \leq s$, then
$$\sum_{i=1}^l q_i \geq \sum_{i=1}^l p_i \ .$$

To see this, let $u := \sum_{i=1}^l p_i$ and assume to the contrary that
$$\sum_{i=1}^l q_i < u \ .$$

Let A be that subset of $\{2^4, 2^5, \cdots\}$ with
$$\sigma(A) = \cup_{i=1}^l E_i \ .$$

It follows from $\sigma^{-1}(n) < \max E_l, \forall n \in \cup_{i=1}^l E_i$, that $2^k < \max E_l, \forall 2^k \in A$. Hence $\max E_l > \max A \geq 2^{u+3}$, since $\overline{\overline{A}} = u$. Thus,
$$q_1 + 1 = \max E_l - \sum_{i=1}^l p_i - \sum_{i=2}^l q_i$$

$$\geq 2^{u+3} - 2u > u + 1 \ .$$

So $q_1 > 0$ and clearly $\sum_{i=1}^l q_i > q_1 > \sum_{i=1}^l p_i$, contradicting our assumption.

Let $l(0) = s + 1$. By the above argument, $\exists$ a largest natural number $1 \leq l(1) < l(0)$ such that

$$\sum_{i=l(1)}^{l(0)-1} q_i \geq \sum_{i=l(1)}^{l(0)-1} p_i \ .$$

Choose $1 <= l(2) < l(1)$ in like fashion, and continue this process to produce $l(0) > l(1) > \cdots > l(a) = 1$. Then $\forall r = 1, 2, \cdots, a$, we have

$$\sum_{i=j}^{l(r-1)-1} q_i < \sum_{i=j}^{l(r-1)-1} p_i, \quad \forall j = l(r) + 1, l(r) + 2, \cdots, l(r-1) - 1,$$

and $\sum_{i=l(r)}^{l(r-1)-1} q_i \geq \sum_{i=l(r)}^{l(r-1)-1} p_i$.

For each $j = 1, 2, \cdots, a$, set

$$M_j := \cup_{i=l(j)}^{l(j-1)-1} E_i \ .$$

Fix j, and write

$$M_j := \left\{ m_1^j < m_2^j < \cdots < m_{k(j)}^j \right\}, \text{ and}$$
$$\cup_{i=l(j)}^{l(j-1)-1} F_i := \left\{ n_1^j < n_2^j < \cdots < n_{e(j)}^j \right\}.$$

Since $\displaystyle\sum_{i=l(j)}^{l(j-1)-1} q_i \geq \sum_{i=l(j)}^{l(j-1)-1} p_i = \overline{\overline{M_j}}$, it follows that $e(j) \geq k(j)$.

Hence we may set $N_j := \left\{ n_{e(j)-k(j)+1}^j, n_{e(j)-k(j)+2}^j, \cdots, n_{e(j)}^j \right\}$, and see that N_j satisfies:

$$\overline{\overline{N_j}} = \overline{\overline{M_j}}, \tag{5}$$

$$N_j \subset \cup_{i=l(j)}^{l(j-1)-1} F_i, \tag{6}$$

$$n_{e(j)-k(j)+i}^j \leq m_i^j, \quad (i = 1, 2, \cdots, k(j)), \text{ and} \tag{7}$$

$$N_i < M_{i-1}, \text{ and } M_i < N_{i-1}, \quad (i = 1, 2, \cdots, a). \tag{8}$$

(Inequality (7) is just Lemma VIII.a.5.)

Since $b_1 \geq b_2 \geq \cdots \geq b_m > 0$, it follows that:

$$\sum \{ b_i : i \in N_j \} > \sum \{ b_i : i \in M_j \}, \quad (j = 1, 2, \cdots, a). \tag{9}$$

The hypotheses of Lemma VIII.a.6 are now satisfied for each sequence of sets

$$\{ E_i : l(j) \leq i < l(j-1) \}, \text{ and}$$
$$\{ F_i : l(j) \leq i < l(j-1) \},$$

and we can conclude that:

$$\left\| \sum \{ b_i t_i : i \in M_j \} \right\| \geq \tfrac{1}{2} \sum \{ |b_i| : i \in M_j \} \\ = \tfrac{1}{2} \sum \{ b_i : i \in M_j \}, \text{ and} \tag{10}$$

$$\| \sum \{b_i t_i : i \in N_j\} \| \geq \tfrac{1}{4} \sum \{|b_i| : i \in N_j\} \tag{11}$$
$$= \tfrac{1}{4} \sum \{|b_i| : i \in N_j\}.$$

By (9), (10), and (11), we have:

$$\| \sum \{b_i t_i : i \in N_j\} \| \geq \tfrac{1}{4} \sum \{b_i : i \in N_j\} \tag{12}$$
$$\geq \tfrac{1}{4} \sum \{b_i : i \in M_j\} \geq \tfrac{1}{4} \| \sum \{b_i t_i : i \in M_j\} \|,$$

for all $j = 1, 2, \cdots, a$.

By the construction,

$$\sum \left\{ b_n t_n : \sigma^{-1(n)} < n \right\} = \sum_{j=1}^{a} \sum \{b_i t_i : i \in M_j\}, \text{ and} \tag{13}$$

$$\exists E \subset \mathbb{N} \text{ such that } \mathrm{E} \left(\sum \left\{ \mathrm{b}_n \mathrm{t}_n : \sigma^{-1}(\mathrm{n}) \geq \mathrm{n} \right\} \right) = \sum_{j=1}^{a} \sum \{\mathrm{b}_i \mathrm{t}_i : i \in \mathrm{N}_j\}. \tag{14}$$

If $u(j) := \min N_j$, and $v(j) := \min E_j$, $(j = 1, 2, \cdots, a)$, then by (8), $u(a) < v(a) < u(a-1) < v(a-1) < \cdots < u(1) < v(1)$. So (12) (together with repeated applications of Proposition I.14 and Corollary II.5) yields

$$
\begin{aligned}
\| \sum \{b_n t_n : \sigma^{-1}(n) < n\} \| &= \| \sum_{j=1}^{a} \sum \{b_i t_i : i \in M_j\} \| \\
&\leq 18 \| \sum_{j=1}^{a} \| \sum \{b_i t_i : i \in M_j\} \| t_{v(j)} \| \\
&\leq 4 \cdot 18 \| \sum_{j=1}^{a} \| \sum \{b_i t_i : i \in N_j\} \| t_{v(j)} \| \\
&\leq 3 \cdot 4 \cdot 18 \| \sum_{j=1}^{a} \| \sum \{b_i t_i : i \in N_j\} \| t_{u(j)} \| \\
&\leq 3 \cdot 4 \cdot 18^2 \| \sum_{j=1}^{a} \sum \{b_i t_i : i \in N_j\} \| \\
&= 2^4 \cdot 3^5 \| E \left(\sum \{b_n t_n : \sigma^{-1}(n) \geq n\} \right) \| \\
&\leq 2^4 \cdot 3^5 \| \sum \{b_n t_n : \sigma^{-1}(n) \geq n\} \|,
\end{aligned}
$$

and this completes the proof of Lemma VIII.a.7. $\square$

Now we are ready for the heart of the matter.

Theorem VIII.a.8: D is a bounded operator on T.

Proof:

Let $x := \sum_n a_n t_n \in T$, and $\{k_1 < k_2 < \cdots\} = \text{supp}(x)$.

Let $Dx := \sum_n b_n t_n$, and $y := \sum_{i=4}^{\infty} |a_{k_{i-3}}| t_{2^i}$.

Note that $Dx = Dy$, and $\|x\| \geq \| \sum a_{k_i} t_i \|$.

By Theorem V.3 and Corollary IV.b.2, $\exists c > 0$ such that for all sequences of scalars $\{d_n\}_{n=1}^{\infty}$, the following holds:

$$\| \sum_{n=4}^{\infty} d_{n-3} t_{2^n} \|_M \leq c \| \sum_{n=1}^{\infty} d_n t_n \|.$$

Finally, let σ be a bijection from $\{4, 5, \cdots\}$ onto $\mathbb{N}$ such that $b_n = a_{k_{\sigma^{-1}(n)-3}}$ induces the nonincreasing rearrangement $Dx = Dy$ of y.

Then by Lemma VIII.a.7 and Remark VIII.a.3,

$$\|Dx\| = \|\textstyle\sum_n b_n t_n\| \; \le \|\textstyle\sum \{b_n t_n : \sigma^{-1}(n) < n\}\| + \|\textstyle\sum \{b_n t_n : \sigma^{-1}(n) \ge n\}\|$$

$$\le [2^4 \cdot 3^5 + 1] \cdot \|\textstyle\sum \{b_n t_n : \sigma^{-1}(n) \ge n\}\|$$

$$\le [2^4 \cdot 3^5 + 1] \cdot \||\textstyle\sum \left\{ a_{k_{\sigma^{-1}(n)-3}} t_{2^{\sigma^{-1}(n)}} : \sigma^{-1}(n) \ge n \right\}\||$$

$$\le [2^4 \cdot 3^5 + 1] \cdot \||\textstyle\sum_{n=4}^{\infty} a_{k_n - 3} t_{2^n}\||$$

$$\le c\,[2^4 \cdot 3^5 + 1] \cdot \|\textstyle\sum_n a_{k_n} t_n\|$$

$$\le c\,[2^4 \cdot 3^5 + 1] \cdot \|x\| \; .$$

□

Notes and Remarks:

1/. It is easy to show that the example given at the beginning of this chapter to show $\|D\| > 1$ is the most economical one. Indeed, for vectors with integer coefficients, D has norm 1 on $[t_i]_{i=1}^6$. Also, any vector x with fewer than 10 non-zero coefficients and with the modulus of each coefficient ≤ 6 satisfies $\|Dx\| = \|x\|$. Finally, any vector x with at most 10 non-trivial coefficients $a_1, a_2, \cdots, a_{10}$ such that for each $i, (1 \le i \le 10), \exists j \ne i, (1 \le j \le 10)$ such that $|a_i - a_j| \le 2$ also satisfies $\|Dx\| = \|x\|$.

2/. The results of this section dualize to T^*. For example, Theorem VIII.a.8 becomes:

Theorem VIII.a.9: $\exists C > 0$ such that for all $x = \sum_n a_n t_n^*$ with $a_1 \ge a_2 \ge \cdots \ge 0$, and any function $f : \mathbb{N} \to \mathbb{N}$, we have

$$\|x\| \ge C \|\textstyle\sum_n a_n t_{f(n)}^*\|.$$

However, D is <u>not</u> a bounded operator on T^*. To see this, let $x_n^* := \sum_{i=n+1}^{2n} t_i^*$. Then for any $x := \sum_n a_n t_n \in T$,

$$|x_n^*(x)| = |\sum_{i=n+1}^{2n} a_i| \le 2\|x\|, \text{ so } \|x_n^*\| \le 2, \forall n \; .$$

On the other hand, for $y := \frac{2}{n} \sum_{i=n+1}^{2n} t_i, \quad \|y\| = 1$, while $x_n^*(y) = 2$, whence $\|x_n^*\| \ge 2$. Thus $\|x_n^*\| = 2$.

But $\|Dx_n^*\| = \|\sum_{i=1}^{n} t_i^*\|$ implies $\lim_n \|Dx_n^*\| = +\infty$, a contradiction.

3/. By the same arguments, if $\{t_{n_i}\}_{i=1}^{\infty}$ is a subsequence of $\{t_n\}_{n=1}^{\infty}$, and for all $x := \sum_i a_i t_{n_i}$ we define

$$Dx = \sum_i b_i t_{n_i},$$

where $\{b_i\}_{i=1}^{\infty}$, is the non-increasing rearrangement of the non-zero elements of $\{|a_i|\}_{i=1}^{\infty}$. then D is a bounded operator. We also clearly have the corresponding result in T^*.

4/. We would hope that there is a more elementary proof of the boundedness of D on T. In particular, it is well known that the following are equivalent for a Banach space X with a basis $\{x_n\}_{n=1}^{\infty}$ and non-increasing rearrangement operator D:

(i) $\|D\| = 1$,

(ii) for any $x = \sum_n a_n x_n \in X$, if $|a_j| < |a_{j+1}|$, for some $j \geq 1$,

then $\|x\| \geq \|\sum_{n=1}^{j-1} a_n x_n + a_{j+1} x_j + a_j x_{j+1} + \sum_{n=j+2}^{\infty} a_n x_n\|$, so if we could find an equivalent norm on T with respect to which $\|D\| = 1$, we would have an elementary proof of the boundedness of D.

5/. The following result is essentially folklore in the area, but there may be some folk that don't know this lore, so we include this result (and a sketch of its proof) for this group of readers.

Theorem VIII.a.10: If $\{x_n\}_{n=1}^{\infty}$ is an unconditional basis for a Banach space X, then the following are equivalent:

(1). D satisfies a "weak triangle law" on X, i.e., there is a constant $K_1 > 0$ such that for all $x, y \in X$,

$$\|D(x + y)\| \leq K_1 (\|Dx\| + \|Dy\|).$$

(2). There is a constant $K_2 > 1$ such that for all scalars $a_1 \geq a_2 \geq \cdots \geq 0$,

$$\|\sum_n a_n x_{2n}\| + \|\sum_n a_n x_{2n-1}\| \leq K_2 \|\sum_n a_n x_n\|.$$

Proof: Without loss of generality, we may assume throughout the proof that the unconditional basis constant of $\{x_n\}_{n=1}^{\infty}$ is 1.

(1) implies (2):

Assume (2) fails. Then for all $K > 1$, there is a decreasing sequence of non-zero scalars $\{a_n\}_{n=1}^{\infty}$ such that

$$\|\sum_n a_n x_{2n}\| + \|\sum_n a_n x_{2n-1}\| > K \|\sum_n a_n x_n\|.$$

If $x = \sum_n a_n x_{2n}, y = \sum_n a_n x_{2n-1}$, and $z = \sum_n a_n x_n$, then $Dx = Dy = Dz = z$, and

$$D(x + y) = x + y \text{ implies}: \quad \begin{aligned} \|Dx\| + \|Dy\| &= 2\|z\| \\ &< \tfrac{2}{K}(\|x\| + \|y\|) \\ &\leq \tfrac{4}{K}\|x + y\| \\ &= \tfrac{4}{K}\|D(x + y)\|, \end{aligned}$$

but if this can be done for each $K > 1$, then D cannot satisfy a weak triangle law.

(2) implies (1):

First we make the following trivial observations:

76

(i) For $x = \sum_n a_n x_n \in X$, write $x(n)$ for a_n. Then if $y \in X$ such that $|x(n)| \le |y(n)|$, $(n = 1, 2, \cdots)$, then $(Dx)(n) \le (Dy)(n)$, for all n.

(ii) If $x, y \in X$, and "v" is defined via

$$x \mathbin{v} y := \sum_n \{|x(n)| \mathbin{v} |y(n)|\} \, x_n,$$

then $|(D(x + y))(n)| \le 2(D(x \mathbin{v} y))(n)$, for all n.

(iii) If $x, y \in X$, and x^2 and $x \cdot y$ are defined via

$$\begin{cases} x^2 = \sum_n x(n) x_{2n} + \sum_n x(n) x_{2n-1}, & \text{and} \\ x \cdot y = \sum_n x(n) x_{2n} + \sum_n y(n) x_{2n-1}, \end{cases}$$

then $|D(x \mathbin{v} y)(n)| \le |(D(x \cdot y))(n)|$, for all n.

(iv) For all $x, y \in X$, and for all n,

$$|(D(x \cdot y))(n)| \le \left|\left(D(x^2)\right)(n)\right| + \left|\left(D(y^2)\right)(n)\right| \ .$$

Now by (i) - (iv), we have for all $x, y \in X$:

$$\begin{aligned} \|D(x + y)\| \ &\le 2\|D(x \mathbin{v} y)\| \\ &\le 2\|D(x \cdot y)\| \\ &\le 2\left(\|Dx^2\| + \|Dy^2\|\right) \le \tfrac{2}{K}\left(\|Dx\| + \|Dy\|\right) \end{aligned}$$

$\square$

It follows that if $\{x_n\}_{n=1}^\infty$ is equivalent to both $\{x_{2n}\}_{n=1}^\infty$ and $\{x_{2n-1}\}_{n=1}^\infty$, then the theorem holds.

Corollary VIII.a.11: D satisfies a weak triangle law on T and T^*.

6/. The lemmas which lead to the boundedness of D on T have analogues which also allow us to prove:

> **Theorem VIII.a.12:** The non-increasing rearrangement operator is a bounded (non-linear) operator on $T^{(2)}$ ("2-convexified Tsirelson space:" see Chapter X.)

7/. We thank T. Barton for stream-lining the proof of Lemma VIII.a.7.

B. Permutations of subsequences of the unit vector basis of Tsirelson's space T.

Since Tsirelson's space contains no subsymmetric sequences, every subsequence $\{t_{n_i}\}_{i=1}^\infty$ of $\{t_n\}_{n=1}^\infty$ has a permutation which is not equivalent to $\{t_{n_i}\}_{i=1}^\infty$. In this section we establish conditions under which $\{t_{n_i}\}_{i=1}^\infty$ is equivalent to a particular permutation of itself. We start with a result which is an immediate consequence of Corollary VII.b.3.

Theorem VIII.b.1: Two subsequences $\{t_{n_i}\}_{i=1}^\infty$ and $\{t_{m_i}\}_{i=1}^\infty$ of $\{t_n\}_{n=1}^\infty$ are permutatively equivalent iff they are equivalent.

Let $\{n_1 < n_2 < \cdots\} \subset \mathbb{N}$ and let σ permute $\{n_i : i \in \mathbb{N}\}$. In general, we may have that $\{t_{\sigma(n_i)}\}_{i=1}^{\infty}$ is not equivalent to $\{t_{n_i}\}_{i=1}^{\infty}$. Let us assume that $\{t_{\sigma(n_i)}\}_{i=1}^{\infty} << \{t_{n_i}\}_{i=1}^{\infty}$. Let

$$I := \{n_i : \sigma(n_i) \leq n_i\} \quad \text{and}$$
$$J := \{n_i : \sigma(n_i) > n_i\} \quad .$$

Then (as observed in the last section), $\exists c > 0$ such that for all scalars $\{b_i\}_{i=1}^{\infty}$,

$$\left\| \sum_{i \in I} b_i t_{\sigma(n_i)} \right\| \leq c \left\| \sum_{i \in I} b_i t_{n_i} \right\| \quad \text{and} \tag{1}$$

$$\left\| \sum_{i \in J} b_i t_{\sigma(n_i)} \right\| \geq c^{-1} \left\| \sum_{i \in J} b_i t_{n_i} \right\|. \tag{2}$$

We will think of σ as mapping the unit vectors $\{t_{n_i} : i \in I\}$ "to the left" and the unit vectors $\{t_{n_i} : i \in J\}$ "to the right". From equations (1) and (2), and the results of Chapter IV, we see that $\{t_{\sigma(n_i)}\}_{i=1}^{\infty}$ is equivalent to $\{t_{n_i}\}_{i=1}^{\infty}$ iff either

a) σ does not move "too many unit vectors too far to the right", or

b) σ does not move "too many unit vectors too far to the left".

Intuitively, one would expect that a permutation of $\{n_i : i \in \mathbb{N}\}$ which moves "many elements far to the left" has left many gaps far to the right, and hence must move "many elements far to the right". In fact, this is precisely what occurs. In other words, $\{t_{\sigma(n_i)}\}_{i=1}^{\infty}$ is equivalent to $\{t_{n_i}\}_{i=1}^{\infty}$ iff either $\{t_{\sigma(n_i)}\}_{i=1}^{\infty} >> \{t_{n_i}\}_{i=1}^{\infty}$, or $\{t_{n_i}\}_{i=1}^{\infty} >> \{t_{\sigma(n_i)}\}_{i=1}^{\infty}$.

To prove this, we need a preliminary result which we choose to prove by appealing to the non-increasing rearrangement operator.

Proposition VIII.b.2: Let $n(1) < n(2) < \cdots$, and $m(1) < m(2) < \cdots$ be natural numbers, and assume that $\{t_{n(i)}\}_{i=1}^{\infty}$ is not equivalent to its subsequence $\{t_{n(m(i))}\}_{i=1}^{\infty}$. Then for each $K > 0, \exists$ a non-increasing sequence $\{b_i\}_{i=1}^{\infty}$ of positive reals such that

$$\left\| \sum_i b_i t_{n(i)} \right\| = 1, \quad \text{and} \quad \left\| \sum_i b_i t_{n(m(i))} \right\| \geq K \quad .$$

Proof: Let D be the non-increasing rearrangement operator given in (3) of the "Notes and remarks" of the last section for the subsequence $\{t_{n(i)}\}_{i=1}^{\infty}$ of $\{t_n\}_{n=1}^{\infty}$. By the proof of Corollary IV.a.2 (see the proof of Theorem IV.a.1), if $\{t_{n(i)}\}_{i=1}^{\infty}$ is not equivalent to $\{t_{n(m(i))}\}_{i=1}^{\infty}$, then $\exists i_0$ and scalars $\{c_i : i_0 < i \leq n(m(i_0))\}$ such that

$$\left\| \sum \left\{ c_i t_{n(i)} : i_0 < i \leq n(m(i_0)) \right\} \right\| = 1 \quad ,$$

while $\left\| \sum \left\{ c_i t_{n(m(i))} : i_0 < i \leq n(m(i_0)) \right\} \right\| \geq 2 \|D\| \cdot K_1$, where $K > 0$ is fixed as in the statement of the proposition, and $K_1 - 6 \geq 5K$.

It follows that for $x := \sum \left\{ c_i t_{n(i)} : i_0 < i \le n(m(i_0)) \right\}$,

$$
\begin{aligned}
2K_1 \|Dx\| \quad &\le 2K_1 \|D\| \cdot \|x\| = 2K_1 \|D\| \\
&\le \sum \left\{ c_i t_{n(m(i))} : i_0 < i \le n(m(i_0)) \right\} \| \\
&\le \sum \{ |c_i| : i_0 < i \le n(m(i_0)) \} \quad .
\end{aligned}
\tag{3}
$$

If we let $\{ \hat{c}_i : i_0 < i \le n(m(i_0)) \}$ be the non-increasing rearrangement of the non-zero elements of $\{ |c_i| : i_0 < i \le n(m(i_0)) \}$, we have by (3),

$$
\begin{aligned}
2K_1 \quad &\| \sum \left\{ \hat{c}_i t_{n(i)} : i_0 < i \le n(m(i_0)) \right\} \| \\
&\le \sum \{ |c_i| : i_0 < i \le n(m(i_0)) \} \\
&= \sum \{ \hat{c}_i : i_0 < i \le n(m(i_0)) \} \\
&\le 2 \| \sum \left\{ \hat{c}_i t_{n(m(i))} : i_0 < i \le n(m(i_0)) \right\} \|.
\end{aligned}
\tag{4}
$$

(This last inequality holds since $\frac{1}{2} \sum \{ \hat{c}_i : i_0 < i \le n(m(i_0)) \}$ is an admissible sum for $\sum \left\{ \hat{c}_i t_{n(m(i))} : i_0 < i \le n(m(i_0)) \right\}$.)

Dividing both sides of (4) by $\| \sum \left\{ \hat{c}_i t_{n(i)} : i_0 < i \le n(m(i_0)) \right\} \|$, we assert the existence of a non-increasing sequence of non-negative scalars $\{a_i\}_{i=i_0+1}^{\infty}$ such that:

$$
\| \sum \left\{ a_i t_{n(i)} : i_0 < i \right\} \| = 1, \text{and}
$$
$$
\| \sum \left\{ a_i t_{n(m(i))} : i_0 < i \right\} \| \ge K_1.
$$

Now define a sequence of scalars $\{d_i\}_{i=1}^{\infty}$ by

$$
d_i := \begin{cases} a_{2i_0}, \text{if } 1 \le i \le 2i_0, \\ a_i, \text{if } 2i_0 \le i \ . \end{cases}
$$

Since $\{a_i\}_{i=i_0+1}^{\infty}$ is non-increasing and non-negative, so is $\{d_i\}_{i=1}^{\infty}$. Thus:

$$
\frac{1}{2} \cdot i_0 a_{2i_0} \le \| \sum_{i=1}^{2i_0} d_i t_{n(i)} \| \le 2i_0 a_{2i_0},
$$

and since $\{a_i\}_{i=1}^{\infty}$ is non-increasing,

$$
\begin{aligned}
i_0 a_{2i_0} \quad &\le \sum \{ a_i : i_0 < i \le 2i_0 \} \\
&\le 2 \| \sum \left\{ a_i t_{n(i)} : i_0 < i \le 2i_0 \right\} \| \le 2.
\end{aligned}
$$

Therefore,

$$
\| \sum_i d_i t_{n(i)} \| \le \| \sum \left\{ d_i t_{n(i)} : 1 \le i \le 2i_0 \right\} \| + \| \sum \left\{ d_i t_{n(i)} : i > 2i_0 \right\} \|
$$
$$
\le 2i_0 a_{2i_0} + 1 \le 5.
$$

On the other hand,

$$
\begin{aligned}
\| \sum_i d_i t_{n(m(i))} \| \quad &\ge \| \sum \left\{ a_i t_{n(m(i))} : i > 2i_0 \right\} \| - 2i_0 a_{2i_0} \\
&\ge \| \sum \left\{ a_i t_{n(m(i))} : i > i_0 \right\} \| - \| \sum \left\{ a_i t_{n(m(i))} : i_0 < i \le 2i_0 \right\} \| - 4 \\
&\ge K_1 - 4 - \sum \{ |a_i| : i_0 < i \le 2i_0 \} \\
&\ge K_1 - 4 - 2 \| \sum \left\{ a_i t_{n(i)} : i_0 < i \le 2i_0 \right\} \| \\
&\ge K_1 - 6 \ge 5K.
\end{aligned}
$$

Now let $b_j := d_j \cdot \| \sum_i d_i t_{n(i)} \|^{-1}$, $(j = 1, 2, \cdots)$. Then $\{b_j\}_{j=1}^\infty$ is a non-increasing sequence of non-negative numbers for which $\| \sum_j b_j t_{n(j)} \| = 1$, and

$$\| \sum_j b_j t_{n(m(j))} \| = \| \sum_j d_j t_{n(m(j))} \| \cdot \| \sum_j d_j t_{n(j)} \|^{-1} > \frac{5K}{5} = K.$$

$\square$

Now we can state the main theorem of this section.

Theorem VIII.b.3: If $\{t_{n(i)}\}_{i=1}^\infty$ is a subsequence of $\{t_n\}_{n=1}^\infty$, $1 \le m(1) < m(2) < \cdots$, and σ permutes $\{n(m(i)) : i \in \mathbb{N}\}$, then the following are equivalent:

(a) $\{t_{\sigma(n(m(i)))}\}_{i=1}^\infty$ is equivalent to $\{t_{n(i)}\}_{i=1}^\infty$.

(b) $\{t_{n(i)}\}_{i=1}^\infty >> \{t_{\sigma(n(m(i)))}\}_{i=1}^\infty$.

Proof: Clearly, (a)$\Rightarrow$(b).

For the other implication, assume that $\{t_{n(i)}\}_{i=1}^\infty$ is not equivalent to $\{t_{\sigma(n(m(i)))}\}_{i=1}^\infty$, but that $\exists K_0 > 0$ such that

$$\| \sum_i a_i t_{\sigma(n(m(i)))} \| \le K_0 \| \sum_i a_i t_{n(i)} \|,$$

for all sequences of scalars $\{a_i\}_{i=1}^\infty$. By Proposition VIII.b.2, $\exists$ a non-increasing sequence of non-negative scalars $\{b_i\}_{i=1}^\infty$ such that $\| \sum_i b_i t_{n(i)} \| = 1$, but $\| \sum_i b_i t_{n(m(i))} \| \ge 2\|D\| \cdot K_0$, where D is the non-increasing rearrangement operator on $[t_{n(m(i))}]_{i=1}^\infty$. Then,

$$\begin{aligned} 2\|D\|K_0 &\le \| \sum_i b_i t_{n(m(i))} \| \\ &\le \|D\| \cdot \| \sum_i b_i t_{\sigma(n(m(i)))} \| \\ &\le K_0 \cdot \|D\| \cdot \| \sum_i b_i t_{n(i)} \| \\ &= K_0 \cdot \|D\|. \end{aligned}$$

This contradiction completes the proof. $\square$

Corollary VIII.b.4: For a subsequence $\{t_{m_i}\}_{i=1}^\infty$ of $\{t_n\}_{n=1}^\infty$ and any permutation σ of $\{m_i : i \in \mathbb{N}\}$, the following are equivalent:

a). $\{t_{m_i}\}_{i=1}^\infty$ is equivalent to $\{t_{\sigma(m_i)}\}_{i=1}^\infty$,

b). $\{t_{m_i}\}_{i=1}^\infty >> \{t_{\sigma(m_i)}\}_{i=1}^\infty$,

c). $\{t_{m_i}\}_{i=1}^\infty << \{t_{\sigma(m_i)}\}_{i=1}^\infty$.

We also obtain:

Corollary VIII.b.5: For subsequences $\{t_{m_i}\}_{i=1}^\infty$ and $\{t_{n_i}\}_{i=1}^\infty$ of $\{t_n\}_{n=1}^\infty$ such that $m_i \le n_i$, $(i = 1, 2, \cdots)$, the following are equivalent:

a) $\{t_{m_i}\}_{i=1}^\infty$ is equivalent to $\{t_{n_i}\}_{i=1}^\infty$,

b) $\{t_{m_{\sigma^{-1}(i)}}\}_{i=1}^{\infty}$ is equivalent to $\{t_{n_i}\}_{i=1}^{\infty}$, for every permutation σ of $\mathbb{N}$ for which $m_{\sigma^{-1}(i)} \leq n_i$, $(i = 1, 2, \cdots)$.

Notes and Remarks:

1/. Proposition VIII.b.2 can be proved by using the proofs of Theorems IV.a.1 and IV.c.1. The proofs of these theorems yield the actual form of the sequence of scalars (as being non-increasing) which makes a subsequence of $\{t_n\}_{n=1}^{\infty}$ not equivalent to a subsequence of itself.

2/. All of the results of this section have straightforward dualizations to T^*.

IX. Unconditional bases for complemented subspaces of Tsirelson's space.

In this chapter we present a series of unpublished results of J. Bourgain, P. Casazza, J. Lindenstrauss, and L. Tzafriri which deal with the complemented subspaces of T that have unconditional bases. Proposition V.11 implies that every normalized basic sequence $\{y_n\}_{n=1}^{\infty}$ in T which is disjointly supported against $\{t_n\}_{n=1}^{\infty}$ has a permutation which dominates $\{t_n\}_{n=1}^{\infty}$. One of the major unsolved problems concerning T is whether or not the hypothesis that the $\{y_n\}_{n=1}^{\infty}$ are disjointly supported is required for this result. As we shall shortly see, this is equivalent to:

Conjecture IX.1: Does T have a unique normalized unconditional basis up to a permutation? Recall the following:

Definition IX.2:

a). A Banach space X with normalized unconditional basis $\{x_n\}_{n=1}^{\infty}$ is said to have a <u>unique normalized unconditional basis up to equivalence</u> if whenever $\{y_n\}_{n=1}^{\infty}$ is a normalized unconditional basis of X the mapping S defined by $Sx_n := y_n, (n = 1, 2, \cdots)$, extends to a linear automorphism of X.

b). A Banach space X with normalized unconditional basis $\{x_n\}_{n=1}^{\infty}$ is said to have a <u>unique normalized unconditional basis up to a permutation</u> ("U.T.A.P.") if, whenever $\{y_n\}_{n=1}^{\infty}$ is a normalized unconditional basis of X, there is a permutation π of $\mathbb{N}$ such that the mapping S defined by $Sx_n := y_{\pi(n)}, (n = 1, 2, \cdots)$, extends to a linear automorphism of X.

It is well known [35] that c_0, l_1, and l_2 are the only Banach spaces which have a unique normalized unconditional basis up to equivalence. M. Edelstein and P. Wojtaszczyk [24] have shown that

$$c_0 \quad \oplus \quad l_1,$$
$$c_0 \quad \oplus \quad l_2,$$
$$l_1 \quad \oplus \quad l_2, \text{and}$$
$$c_0 \quad \oplus \quad l_1 \oplus l_2 \quad \text{each have U.T.A.P.}$$

[15] made this list a good deal longer by showing that

$$\Sigma \left(l_1^k \oplus l_2^m \oplus l_\infty^n \right)_1, (k, m, n = 1, 2, \cdots),$$
$$\left(\sum_n \oplus l_k^n \right)_0, (k = 1, 2),$$
$$l_k \oplus \left(\sum_n \oplus l_k^n \right)_0, (k = \infty, 1, 2),$$
$$(l_k \oplus l_k \oplus \cdots)_0, (k = 1, 2),$$
$$(c_0 \oplus c_0 \oplus \cdots)_1, \text{ (and a few more such spaces)}$$

<u>all</u> have U.T.A.P. We have not resolved Conjecture IX.1, but in Chapter X we will demonstrate that the 2-convexification of T has U.T.A.P., this being the first such space known not "built"

out of the classical sequence spaces.

One indication that unconditional bases for complemented subspaces of T might behave similarly to permutations of subsequences of $\{t_n\}_{n=1}^{\infty}$ is Theorem VII.b.4, which asserts that a normalized K-unconditional basis $\{x_i\}_{i=1}^{n}$ for a $\|P\|$-complemented subspace of T has a "percentage" of the basis which is equivalent to the unit vector basis of l_1^n, with the equivalence constant being a function of $K, \|P\|$, and the percentage desired. This property is basic to all subsequences of $\{t_n\}_{n=1}^{\infty}$, and Theorem VII.b.4 shows that it is inherited by unconditional bases for complemented subspaces of T.

Our initial result is a strong property held by all unconditional basis of T, and appears here for the first time.

Theorem IX.3: Every normalized unconditional basis for T has a subsequence which is permutatively equivalent to $\{t_n\}_{n=1}^{\infty}$.

Although we have already covered the basic notions required to prove the above, we need to refine some of them before we give the proof. Our first lemma states the (almost) apparent fact that if we delete from $\{t_i\}_{i=1}^{m}$ a subsequence which is equivalent (up to a fixed constant) to the unit vector basis of l_1^k, then the equivalence constant between the remaining vectors and the unit vector basis of $l_1^{m-n-k+1}$ is approximately the same as the equivalence constant between $\{t_i\}_{i=1}^{m}$ and the unit vector basis of l_1^{m-n+1}.

Lemma IX.4: For each j, $\exists c := c(j)$ such that for every $n < m$, whenever $E \subset \{n, n+1, \cdots, m\}$ satisfies $\overline{\overline{E}} \geq m - \exp_j(n) > 0$, then

$$d_0\left([t_i]_{i=n}^{m}, l_1^{m-n+1}\right) \leq d_0\left([t_i : i \in E], l_1^{\overline{\overline{E}}}\right) + c(j),$$

where d_0 is the canonical distance between the unit vector bases of the indicated spaces (i.e., for $E \subset \mathbb{N}, d_0\left([t_i : i \in E], \ell_1^{\overline{\overline{E}}}\right) = \|I\|$, where I is the formal identity from $[t_i : i \in E]$ to $\ell_1^{\overline{\overline{E}}}$.)

Proof: Fix $j \in \mathbb{N}$, and let $E \subset \{n, n+1, \cdots, m\}$ such that $\overline{\overline{E}} \geq m - \exp_j(n)$. If $F = \{i : n \leq i \leq m, \text{ and } i \notin E\}$, then $\overline{\overline{F}} \leq \exp_j(n)$. Let

$$X := [t_i : i \in E], \text{ and } Y := [t_i : i \in F].$$

By Remark 4 of Chapter IV.B (Notes and remarks), $\exists$ an isomorphism L_0 on T defined by

$$L_0(t_i) := t_{\exp_j(i)}, (i = 1, 2, \cdots).$$

We now define $c(j) := 2\|L_0\| \cdot \|L_0^{-1}\|$. Since $\exp_j(i) \geq \exp_j(n), \forall i \in F$, and since $\overline{\overline{F}} \leq \exp_{j(i)}$, it follows that $\left\{t_{\exp_{j(i)}} : i \in F\right\}$ is 2-equivalent to the unit vector basis of $l_1^{\overline{\overline{F}}}$. Thus

$$d_0\left([t_i]_{i=n}^{m}, l_1^{m-n+1}\right) \leq$$
$$d_0\left([t_i : i \in E], l_1^{\overline{\overline{E}}}\right) + d_0\left([t_i : i \in F], l_1^{\overline{\overline{F}}}\right) \leq$$
$$d_0\left([t_i : i \in E], l_1^{\overline{\overline{E}}}\right) + 2\|L_0\| \cdot \|L_0^{-1}\| =$$
$$d_0\left([t_i : i \in E], l_1^{\overline{\overline{E}}}\right) + c(j).$$

Our next lemma should be considered in the following context:

Let $\{x_i\}_{i=1}^{\infty}$ be a K-unconditional basis for T, and let $\{x_i^*\}_{i=1}^{\infty}$ be the sequence of associated coefficient functionals. As before, let

$$E_{n,m} := \{n, n+1, \cdots, m\}, \text{ and}$$
$$Y_{n,m} := [t_i]_{i=n}^{m}, (\text{ for } n < m).$$

If $|x_i^*(E_{n,m}x_i)| < \frac{1}{2}$, for fixed $E_{n,m}$ and all $i \in \mathbb{N}$, then by Lemma VII.a.2, $\exists$ an isomorphism of $Y_{n,m}$ onto a complemented subspace of $X \oplus Y$, where

$$X := (X_1 \oplus X_1 \oplus \cdots)_{l_1}, \text{ and}$$
$$Y := (Y_1 \oplus Y_1 \oplus \cdots)_{l_1}, \text{ where}$$
$$X_1 := [t_i]_{i=1}^{n} - 1, \text{ and } Y_1 := [t_i]_{m+1}^{\infty}.$$

By Lemma VII.a.1, "most" of the t_i (where $n \leq i \leq m$) will map to elements with "most" of their support in Y. Now (by invoking Lemmas VII.a.2 and IX.4) a portion of those t_i in $Y_{n,m}$ span a subspace M whose canonical distance to l_1^k is the same as the canonical distance of $Y_{n,m}$ to l_1^{m-n+1}, and moreover M is isomorphic to a complemented subspace of $(\sum \oplus Y)_{l_1}$. But by Lemma VII.a.3, this can only occur if $\{t_i\}_{i=n}^{m}$ is equivalent to the unit vector basis of l_1^{m-n+1}. Since each result used in this argument is quantitative, we should be able to delimit this distance, and we do so in the following lemma:

Lemma IX.5: Let $\{x_i\}_{i=1}^{\infty}$ be a normalized K-unconditional basis for T, and let $\{x_i^*\}_{i=1}^{\infty}$ be its sequence of associated coefficient functionals. Letting

$$E_{n,m} := \{n, n+1, \cdots, m\}, \text{ and}$$
$$Y_{n,m} := [t_i]_{i=n}^{m},$$

$\exists$ a constant $M (:= M(K))$ so that for any $n < m$, if $|x_i^*(E_{n,m}x_i)| < \frac{1}{2}$, for $i = 1, 2, \cdots$, then $d_0\left(Y_{n,m}, l_1^{\overline{E}_{n,m}}\right) \leq M$, (where d_0 is defined as a Lemma IX.4).

Proof: First choose the natural number $j := j\left(K, K, \frac{1}{4}\right)$ given in Lemma VII.a.1. Next, choose a natural number $k := k(j)$ such that

$$\exp_j\left(\exp_j(n)\right) \leq \exp_k(n) - n, (n = 1, 2, \cdots).$$

Finally let $c := c(k)$ be the constant given in Lemma IX.4. Fix $n < m$. Since $\{x_i\}_{i=1}^{\infty}$ is a normalized K-unconditional basis for T, $\exists\ q \in \mathbb{N}$ such that $Y_{n,m}$ is 2-isomorphic to a complemented subspace of $[x_i]_{i=1}^{q}$. Letting

$$\text{set} \quad \begin{cases} E_n := \{1, 2, \cdots, n-1\}, \text{ and } F_m := \{m, m+1, \cdots\}, \\ F := \left\{i : |x_i^*(F_m x_i)| \geq \frac{1}{4}, 1 \leq i \leq q\right\}, \text{ and} \\ E := \{i : 1 \leq i \leq q, i \notin F\}. \end{cases}$$

For $i \in E : |x_i^*(F_m x_i)| < \frac{1}{4}$, while $|x_i^*(E_{n,m} x_i)| < \frac{1}{2}$.

Thus $|x_i^*(E_n x_i)| \geq \frac{1}{4}$. By Lemma VII.a.2, $[x_i : i \in E]$ is $M_0 := M_0(K)$-isomorphic to some M_0-complemented subspace, say X, of

$$\left(\sum_{s=1}^{\infty} \oplus [t_i : i \in E_s] \right)_{l_1} .$$

In like fashion, $[x_i : i \in F]$ is M_0-isomorphic to some M_0-complemented subspace, say Y, of

$$\left(\sum_{s=1}^{\infty} \oplus [t_i : i \in F_s] \right)_{l_1} .$$

By Lemma VII.a.1,

$$\dim X = \overline{\overline{E}} \leq \exp_j(n).$$

It follows that $[t_i : i \in E_{n,m}]$ is $2M_0$-isomorphic to a $2M_0$-complemented subspace of $X \oplus Y$. Let $\{z_i\}_{i=n}^m$ be the set of normalized images in $X \oplus Y$ (under this isomorphism) of the set of vectors $\{t_i\}_{i=n}^m$. Let Q be the natural projection of $X \oplus Y$ onto $[z_i : i \in E_{n,m}]$. Let $z_i = x_i + y_i$ be the decomposition of z_i against X and Y, for all $i \in E_{n,m}$, and let $\{z_i^* : i \in E_{n,m}\}$ consist of linear functionals on $X \oplus Y$ for which

$$Qz = \sum_{i=n}^m z_i^*(z)z_i, \quad \forall z \in X \oplus Y.$$

Now let
$$F_0 := \left\{ i : i \in E_{n,m}, |z_i^*(y_i)| \geq \frac{1}{2} \right\}, \text{ and}$$
$$E_0 := E_{n,m} \sim F_0.$$

Applying Lemma VII.a.1 again, we have

$$\overline{\overline{E_0}} \leq \exp_j(\dim X) = \exp_j(\overline{\overline{E}})$$
$$\leq \exp_j(\exp_j(n)) \leq \exp_k(n) - n.$$

Thus $\overline{\overline{F_0}} \geq m - \exp_k(n)$. Since $\overline{\overline{F_0}} \leq m$, it follows by Lemmas VII.a.2 and VII.a.3 that $\{t_i : i \in F_0\}$ is $f(M_0, 2M_0)$-equivalent to the unit vector basis of $l_1^{\overline{\overline{F_0}}}$. Therefore by Lemma IX.4, we obtain

$$d_0\left(Y_{n,m}, l_1^{\overline{\overline{E}}_{n,m}}\right) \leq$$
$$d_0\left([t_i : i \in F_0], l_1^{\overline{\overline{F_0}}}\right) + c(k) \leq$$
$$f(M_0, 2M_0) + c(k) =: M.$$

Since M_0 and k are functions of K, we are done. $\square$

Using these lemmas we can now proceed with:

Proof of Theorem IX.9: Let $\{x_i\}_{i=1}^{\infty}$ be a normalized K-unconditional basis for T with associated coefficient functionals $\{x_i^*\}_{i=1}^{\infty}$. Let M be given by Lemma IX.5 and choose

$$n_0 = 1 < n_1 < n_2 < \cdots$$

such that if $\alpha_j := d_0\left(Y_{n_j+1,n_{j+1}}, l_1^{n_{j+1}-n_j}\right), (j = 1, 2, \cdots)$, where

$$Y_{n,m} := [t_i]_{i=n}^m, (n < m), \text{ then}$$

$$M < \alpha_j \le M + 1, \quad (j = 1, 2, \cdots). \tag{1}$$

By Lemma VII.a.2, for each $j, \exists i(j) \in \mathbb{N}$ such that

$$\left| x^*_{i(j)} \left(E_{n_j, n_j+1} x_{i(j)} \right) \right| \ge \frac{1}{2}, \quad \text{where} \tag{2}$$

$$E_{n,m} := \{n, n+1, \cdots, m\}, \quad \text{for all } n < m.$$

Since the indices $i(j)$ may not be distinct, we will thin them out by deleting any $i(k)$ such that

$$i(k) = i(j), \quad \text{for some } j < k.$$

Let E be the set of all (distinct) indices j which remain after this "thinning". Now define
$P : T \to \left[x_{i(j)} : j \in E \right]$ by $P \left(\sum_i a_i x_i \right) := \sum_{j \in E} a_i(j) x_{i(j)}$, for all scalar sequences $\{a_i\}_{i=1}^\infty$.
Since $\{x_i\}_{i=1}^\infty$ is an unconditional basis for T, P is a bounded linear projection on T.

By (2), the diagonal operator Δ of the operator

$$P\big|_{\left[E_{n_j, n_j+1} x_{i(j)} : j \in E \right]}$$

satisfies

$$\|\Delta\| \le 2\|P\|.$$

We have then, for all scalar sequences $\{a_j\}_{j=1}^\infty$,

$$\|\sum_{j \in E} a_j x_{i(j)}\| \le 2\|P\| \cdot \|\sum_{j \in E} a_j E_{n_j, n_j+1} x_{i(j)}\|.$$

Thus by invoking Proposition II.4,

$$\|\sum_{j \in E} a_j x_{i(j)}\| \le 36\|P\| \cdot \|\sum_{j \in E} a_j t_{n_j}\|.$$

Conversely by Theorem I.d.6 of Volume II [36], $\exists \alpha > 0$ such that for all scalar sequences $\{a_j\}_{j \in E}$,

$$\|\sum_{j \in E} a_j x_{i(j)}\| \ge \alpha\| \left(\sum_{j \in E} \left| a_j x_{i(j)} \right|^2 \right)^{\frac{1}{2}} \|,$$

where $|\cdot|$ is taken relative to the lattice structure on T generated by $\{t_n\}_{n=1}^\infty$.

Hence

$$\|\sum_{j \in E} a_j x_{i(j)}\| \ge \|\alpha \left(\sum_{j \in E} \left| a_j E_{n_j, n_j+1} x_{i(j)} \right|^2 \right)^{\frac{1}{2}} \|$$
$$= \alpha\|\sum_{j \in E} a_j E_{n_j, n_j+1} x_{i(j)}\|.$$

Thus by Corollary II.5,

$$I\|\sum_{j \in E} a_j x_{i(j)}\| \ge \frac{\alpha}{6}\|\sum_{j \in E} a_j t_{n_j}\|,$$

i.e., $\left\{ x_{i(j)} : j \in E \right\}$ is equivalent to $\left\{ t_{n_j} : j \in E \right\}$. We will now show that this latter set (when indexed in the natural order) is equivalent to $\{t_j\}_{j=1}^\infty$.

Let $n_j < n_h$ be two "consecutive" n_j's, for $i \in E$, namely: if $n_j < n_k < n_h$, then $k \notin E$. Let $\delta := \{n_k : n_j < n_k < n_h\}$. If $n_k \in \delta$, then $k \notin E$, and there must exist an l such that

$$n_l < n_j, \text{ and } i(l) = i(k).$$

Thus $\overline{\overline{\delta}} \le n_j$ and $\{t_{n_k} : n_k \in \delta\}$ is 2-equivalent to the unit vector basis of $l_1^{\overline{\overline{\delta}}}$. By the remark following Theorem IV.a.1, we conclude that $\{t_{n_j} : j \in E\}$ (when indexed in the natural order) is equivalent to $\{t_{n_j}\}_{j=1}^\infty$. Finally note that, by our choice of n_j, (1) holds. Hence by Theorem IV.a.1, $\{t_{n_j}\}_{j=1}^\infty$ is equivalent to $\{t_j\}_{j=1}^\infty$, and thus $\{x_{i(j)} : j \in E\}$ is equivalent to $\{t_j\}_{j=1}^\infty$. □

As indicated earlier, Proposition V.11 implies that every normalized disjointly supported basic sequence $\{y_n\}_{n=1}^\infty$ in T has a permutation which dominates $\{t_n\}_{n=1}^\infty$. On the other hand, our next theorem will show that $\{t_n\}_{n=1}^\infty$ dominates (up to a permutation) every normalized unconditional basis for T. To demonstrate this, we first require a strengthening of Proposition V.11 for non-disjointly supported vectors.

Lemma IX.6: If $\{y_n\}_{n=1}^\infty$ is a normalized sequence of vectors in T such that $y_n \in [t_i : l_n \le i < l_{n+1}]$ for all n and for some $l_1 < l_2 < \cdots$, and if

$$A_i := \{n : E_i y_n \ne 0\}, \quad \forall i,$$

where $E_i := \{1, 2, \cdots, i\}$, then $\exists M > 0$ such that for all sequences of scalars $\{a_n\}_{n=1}^\infty$, we have:

$$\sup_{\theta_n = \pm 1} \|\sum_n \theta_n a_n y_n\| \le M \|\sum_n a_n t_{l_n}\|.$$

Proof: Let $\||| \cdot \|||$ be the "modified" form of the f-Tsirelson norm defined for $f(k) := 2k + 2$, as in Remark 3 of Chapter I, i.e. $\||| \cdot \|||$ sums $2k + 2$ disjointly supported vectors with support beyond $k - 1$. By Corollary IV.b.2 and Theorem V.3, we know that $\||| \cdot \|||$ is equivalent to the usual Tsirelson norm. Hence it suffices to show:

(1). For all sequences of scalars $\{a_n\}_{n=1}^\infty$, all sequences of signs $\{\theta_n\}_{n=1}^\infty$, and all $m = 0, 1, 2, \cdots$, that:

$$\|\sum_n \theta_n a_n \frac{y_n}{4}\|_m \le \||| \sum_n a_n t_{l_n} \|||_m.$$

We proceed by induction, with the case $m = 0$ clearly holding. Assume that (1) holds for some $m \ge 0$, and for all choices of scalars $\{a_n\}_{n=1}^\infty$ and signs $\{\theta_n\}_{n=1}^\infty$.

Let $x := \sum_n \theta_n a_n \cdot \frac{y_n}{4}$, for some scalars $\{a_n\}_{n=1}^\infty$ and signs $\{\theta_n\}_{n=1}^\infty$. Choose finite subsets of $\mathbb{N}$:

$$k \le G_1 < G_2 < \cdots < G_k.$$

Since the vectors y_i have finite support and since

$$\overline{\overline{A_i}} \le i, \quad \forall i,$$

$\exists$ a set $G_{k+1} > G_k$ such that if $G_i y_n \neq 0$, for some $1 \leq i \leq k$, and some n, then

$$\max \ \mathrm{supp} \ y_n \leq \max G_{k+1}.$$

For each $1 \leq j \leq k$, let

$$F_j := \{n : G_j y_n \neq 0, \ \text{but} \ G_s y_n = 0, \ \text{for all} \ s > j\}.$$

For each $1 \leq j \leq k+1$, let

$$
\begin{aligned}
F_j^1 &:= \{n \in F_j : G_j y_n = y_n\}, \ \text{and} \\
F_j^2 &:= F_j \sim F_j^1.
\end{aligned}
$$

By triangulation and the definitions of the sets F_j^i, we have:

$$
\begin{aligned}
\tfrac{1}{2}\sum_{j=1}^{k}\|G_j x\|_m &\leq \tfrac{1}{2}\sum_{j=1}^{k+1}\sum_{i=1}^{j}\|G_i \sum\left\{\theta_n a_n \tfrac{y_n}{4} : n \in F_j\right\}\|_m \\
&\leq \tfrac{1}{2}\sum_{j=1}^{k+1}\|\sum\left\{\theta_n a_n \tfrac{y_n}{4} : n \in F_j^1\right\}\|_m + \\
&\quad \tfrac{1}{2}\sum_{j=1}^{k+1}\sum_{i=1}^{j}\|G_i \sum\left\{\theta_n a_n \tfrac{y_n}{4} : n \in F_j^2\right\}\|_m \\
&\leq \tfrac{1}{2}\sum_{j=1}^{k+1}\|\sum\left\{\theta_n a_n \tfrac{y_n}{4} : n \in F_j^1\right\}\|_m + \\
&\quad \tfrac{1}{2}\sum_{j=1}^{k+1}\sum\left\{\sum_{i=1}^{j}\|\theta_n a_n G_1 \tfrac{y_n}{4}\|_m : n \in F_j^2\right\}.
\end{aligned}
$$

Upon applying the inductive hypothesis to each summand in the first sum above and the definition of the norm to the second sum, we obtain:

$$
\begin{aligned}
\text{(2).} \quad \tfrac{1}{2}\sum_{j=1}^{k}\|G_j x\|_m &\leq \tfrac{1}{2}\sum_{j=1}^{k+1}\||\sum\left\{a_n t_{l_n} : n \in F_j^1\right\}\||_m + \\
&\quad \tfrac{1}{2}\sum_{j=1}^{k+1}\sum\left\{2\|\theta_n a_n \cdot \tfrac{y_n}{4}\|_m : n \in F_j^2\right\} \\
&\leq \tfrac{1}{2}\sum_{j=1}^{k+1}\||\sum\left\{a_n t_{l_n} : n \in F_j^1\right\}\||_m + \\
&\quad \tfrac{1}{4}\sum_{j=1}^{k+1}\sum\left\{|a_n| : n \in F_j^2\right\}.
\end{aligned}
$$

Let $s_j := \min G_j, (j = 1, 2, \cdots, k)$, and note that $n \in F_j^2$ implies $G_j y_n \neq 0$, but $G_s y_n = 0$, for all $s > j$.

Thus $n \in A_{s_j}$, and by hypothesis,

$$\overline{\overline{F_j^2}} \leq \overline{\overline{A_{s_j}}} \leq s_j, \quad (j = 1, 2, \cdots, k).$$

Therefore:

$$\text{(3).} \quad \tfrac{1}{2}\sum\left\{|a_n| : n \in F_j^2\right\} \leq \||\sum\left\{a_n t_{l_n} : n \in F_j^2\right\}\||_m, \quad (j = 1, 2, \cdots, k).$$

Combining (2), (3), and the definition of $||| \cdot |||_m$, we obtain:

$$\frac{1}{2}\sum_{j=1}^{k}\|E_j x\|_m \leq \frac{1}{2}\sum_{j=1}^{k+1}||| \sum \left\{a_n t_{l_n} : n \in F_j^1\right\}|||_m +$$

$$\frac{1}{2}\sum_{j=1}^{k+1}||| \sum \left\{a_n t_{l_n} : n \in F_j^2\right\}|||_m$$

$$\leq ||| \sum_{j=1}^{k+1} \sum \{a_n t_{l_n} : n \in F_j\}|||_{m+1}$$

$$\leq ||| \sum_n a_n t_{l_n}|||_{m+1}.$$

By the definition of the usual norm on T, we now have

$$\| \sum_n \theta_n a_n \frac{y_n}{4}\|_{m+1} \leq ||| \sum_n a_n t_{l_n}|||_{m+1},$$

for all scalars $\{a_n\}_{n=1}^\infty$ and all signs $\{\theta_n\}_{n=1}^\infty$, i.e. (1) for m implies (1) for $m+1$. □

The next theorem is an unpublished result of P. Casazza, and will be proven in gradual steps due to its technicality.

Theorem IX.7: If $\{x_n\}_{n=1}^\infty$ is a normalized unconditional basis for T, then $\exists$ a permutation σ of $\mathbb{N}$ such that:

$$\{x_{\sigma(n)}\}_{n=1}^\infty << \{t_n\}_{n=1}^\infty.$$

Proof:

Step I: the construction:

As we have done before, we let

$$E_n := \{1, 2, \cdots, n\},$$
$$E_{n,m} := \{n, n+1, \cdots, m\}, \quad (\text{for } n < m), \quad \text{and}$$
$$E_n' := \{n, n+1, \cdots\}.$$

Let $\{x_n\}_{n=1}^\infty$ be a normalized unconditional basis for T, with $\{x_n^*\}_{n=1}^\infty$ the associated coefficient functionals. For each m, choose j_m to be the largest natural number such that

$$x_m^*\left(E_{j_m}' x_m\right) \geq \frac{1}{4}. \tag{1}$$

It follows immediately that

$$x_m^*\left(E_{j_m+1}' x_m\right) < \frac{1}{4}, \tag{2}$$

and that

$$x_m^*\left(E_{j_m} x_m\right) \geq \frac{3}{4}. \tag{3}$$

For each m, let $w_m := E_{j_m}' x_m$, and choose i_m so that

$$x_m^*\left(E_{i_m}' w_m\right) \geq \frac{1}{8}, \tag{4}$$

and

$$x_m^*\left(E_{i_m} w_m\right) \geq \frac{1}{8}. \tag{5}$$

89

Now let

$$y_m := E_{i_m} w_m, \tag{6}$$

and

$$z_m := E'_{i_m} w_m. \tag{7}$$

Note that supp $y_m \cap$ supp $z_m = \{i_m\}, (m = 1, 2, \cdots)$. Without loss of generality, we may assume that we have permuted $\{x_m\}_{m=1}^\infty$ so that $i_1 \le i_2 \le \cdots$. Now divide $\mathbb{N}$ into non-void sets $\{F_j\}_{j=1}^\infty$ such that

$$F_1 < F_2 < \cdots, \tag{8}$$

and

$$m_1, m_2 \in F_j \text{ implies } i_{m_1} = i_{m_2}, \tag{9}$$

and

$$m_1 \in F_j \text{ and } m_2 \in F_k \text{ with } j \ne k \text{ implies } i_{m_1} \ne i_{m_2}. \tag{10}$$

For each j, let $k_j := \overline{\overline{F}}_j$, and select a representative

$$m(j) \in F_j.$$

Let H be the space of all $x := \sum_j x_j$, where $x_j \in l_1^{k_j}$, and H is normed by

$$\|x\| := \|\sum_j \|x_j\|_{l_1^k}, t_{i_m(j)}\|_T,$$

and H consists solely of those such x for which this norm is finite. Let $\{h_n\}_{n=1}^\infty$ be the natural unit vector basis for H. Finally, define sequences $\{y'_m\}_{m=1}^\infty$ and $\{z'_m\}_{m=1}^\infty$ by

$$\|\sum_m a_m y'_m\| = \sup_{\theta_m = \pm 1} \|\sum_m \theta_m a_m y_m\|_T, \tag{11}$$

and

$$\|\sum_m a_m z'_m\| = \sup_{\theta_m = \pm 1} \|\sum_m \theta_m a_m z_m\|_T, \tag{12}$$

for all choices of scalars $\{a_n\}_{n=1}^\infty$.

We will use the auxiliary space H as follows:

Step II: $\{h_n\}_{n=1}^\infty$ is equivalent to a subsequence of $\{t_n\}_{n=1}^\infty$.

By Lemma VII.a.1, $\exists$ a constant $C := j\left(K, K, \frac{1}{8}\right)$, where K is the unconditionality constant for the basis $\{x_n\}_{n=1}^\infty$, so that

$$\sum_{j=1}^q k_j \le \exp_C(i_{m(q)}), \text{ for all } q = 1, 2, \cdots.$$

Choose $k'_j \ge k_j, (j = 1, 2, \cdots)$, such that

$$\sum_{j=1}^q k'_j \le \exp_C(i_{m(q)}), \text{ for all } q = 1, 2, \cdots.$$

90

Define H' just as H was defined, but replacing k_j by k'_j throughout.

Since $k'_j \geq k_j, \forall j$, H occurs naturally as a subspace of H', and its basis $\{h_n\}_{n=1}^\infty$ occurs naturally as a subsequence of the unit vector basis of H'.

So $\{h_n\}_{n=1}^\infty$ will be equivalent to a subsequence of $\{t_n\}_{n=1}^\infty$ if the unit vector basis of H' is equivalent to a subsequence of $\{t_n\}_{n=1}^\infty$. To prove this latter claim note that if $K_1 > 0$ is the constant guaranteed by (3) of the "Notes and Remarks" (Chapter IV), and if we apply Proposition II.4, then for any sequence of scalars $\left\{a_{j,n} : j = 1, 2, \cdots; n = 1, 2, \cdots, k'_j\right\}$, we obtain:

$$
\begin{aligned}
&\|\sum_j \sum_{n=1}^{k'_j} a_{j,n} t_{n+\exp_C(i_{m(j)})}\| \leq \\
&18\|\sum_j \|\sum_{n=1}^{k'_j} a_{j,n} t_{n+\exp_C(i_{m(j)})}\| t_{\exp_C(i_{m(j)})}\| \leq \\
&18\|\sum_j \left(\sum_{n=1}^{k'_j} |a_{j,n}|\right) t_{\exp_C(i_{m(j)})}\| \leq \\
&18K_1\|\sum_j \left(\sum_{n=1}^{k'_j} |a_{j,n}|\right) t_{i_{m(j)}}\|.
\end{aligned}
\tag{13}
$$

Similarly we can find a constant $K_2 > 0$ such that

$$
\begin{aligned}
&\|\sum_j \left(\sum_{n=1}^{k'_j} |a_{j,n}|\right) t_{i_{m(j)}}\| \leq \\
&K_2\|\sum_j \|\sum_{n=1}^{k'_j} a_{j,n} t_{n+\exp_C(i_{m(j)})}\| t_{\exp_C(i_{m(j)})}\| \leq \\
&3K_2\|\sum_j \|\sum_{n=1}^{k'_j} a_{j,n} t_{n+\exp_C(i_{m(j)})}\|,
\end{aligned}
\tag{14}
$$

for all scalar choices $\left\{a_{j,n} : j = 1, 2, \cdots; n = 1, 2, \cdots, k'_j\right\}$. If $\left\{h'_{j,n} : j = 1, 2, \cdots; n = 1, 2, \cdots, k'_j\right\}$ is the unit vector basis of H', then (13) and (14) imply that $\left\{h'_{j,n} : j = 1, 2, \cdots; n = 1, 2, \cdots, k'_j\right\}$ is equivalent to the subsequence $\left\{t_{n+\exp_C(i_{m(j)})} : j = 1, 2, \cdots; \text{ and } n = 1, 2, \cdots, k'_j\right\}$ of $\{t_n\}_{n=1}^\infty$.

Step III: $\{x_n\}_{n=1}^\infty << \{y'_n\}_{n=1}^\infty$.

It follows from (2), (4), and (6) that

$$
x_n^*(y_n) \geq \frac{1}{4} - \frac{1}{8} = \frac{1}{8}, \quad (n = 1, 2, \cdots).
$$

Thus by Remark 2, p. 21, [35], $\{x_n\}_{n=1}^\infty << \{y'_n\}_{n=1}^\infty$.

Step IV: $\{y'_n\}_{n=1}^\infty << \{h_n\}_{n=1}^\infty$.

Let L be that isomorphism on T whose existence is guaranteed by (3) of the "Notes and Remarks" (Chapter IV) which is defined by

$$
Lt_n = t_{\exp_C(n)}, \quad (n = 1, 2, \cdots).
$$

Define $(Ly_n)'$ by:

$$\left\| \sum_n a_n (Ly_n)' \right\| := \sup_{\theta_n = \pm 1} \left\| \sum_n \theta_n a_n (Ly_n) \right\|_T,$$

for all scalar choices $\{a_n\}_{n=1}^\infty$.

Clearly $\{y_n'\}_{n=1}^\infty$ as defined in (11) is equivalent to $\{(Ly_n)'\}_{n=1}^\infty$ as defined above. It now suffices to prove that

$$\{(Ly_n)'\}_{n=1}^\infty << \left\{ t_{n + \exp_C(i_{m(j)})} : j = 1, 2, \cdots; \text{ and } n = 1, 2, \cdots, k_j \right\}.$$

For any n, $x_n^*(y_n) \geq \frac{1}{8}$, if $a_n := \{i : E_n y_i \neq 0\}$, then by Lemma VII.a.2,

$$\overline{\overline{A_n}} \leq \exp_C(n).$$

Hence if $B_n := \{i : E_n(Ly_i) \neq 0\}$, then $\overline{\overline{B_n}} \leq n$. Step IV now follows immediately from Lemma IX.6.

Step IV: $\exists$ a permutation π of $\mathbb{N}$ such that $\{x_{\pi(n)}\}_{n=1}^\infty << \{t_n\}_{n=1}^\infty$.

By I-IV, $\exists$ a permutation π of $\mathbb{N}$ and a subsequence $\{t_{n_i}\}_{i=1}^\infty$ of $\{t_n\}_{n=1}^\infty$ such that

$$\{x_{\pi(i)}\}_{i=1}^\infty << \{t_{n_i}\}_{i=1}^\infty.$$

By Theorem IX.3, $\exists$ a permutation σ of $\mathbb{N}$ and a subsequence $\{x_{\pi(i(j))}\}_{j=1}^\infty$ of $\{x_{\pi(i)}\}_{i=1}^\infty$ such that

$$\{x_{\pi(i(j))}\}_{j=1}^\infty \text{ is equivalent to } \{t_{\sigma(j)}\}_{j=1}^\infty.$$

It follows that

$$\{t_{n_{i(j)}}\}_{j=1}^\infty >> \{t_{\sigma(j)}\}_{j=1}^\infty.$$

By Theorem VIII.b.3, it follows that

$$\{t_{n_{i(j)}}\}_{j=1}^\infty \text{ is equivalent to } \{t_j\}_{j=1}^\infty.$$

Thus $\{t_{n_i}\}_{i=1}^\infty$ is equivalent to $\{t_i\}_{i=1}^\infty$, and so

$$\{x_{\pi_i}\}_{i=1}^\infty << \{t_i\}_{i=1}^\infty.$$

$\square$

Corollary IX.8: If $\{y_n\}_{n=1}^\infty$ is a sequence of normalized vectors which are disjointly supported and $[y_n]_{n=1}^\infty$ is isomorphic to T, then $\{y_n\}_{n=1}^\infty$ is permutatively equivalent to $\{t_n\}_{n=1}^\infty$.

Proof: By Proposition V.11, $\exists$ a permutation σ of $\mathbb{N}$ such that

$$\{t_n\}_{n=1}^\infty << \{y_{\sigma(n)}\}_{n=1}^\infty.$$

By Theorem IX.7, $\exists$ a permutation π of $\mathbb{N}$ such that

$$\{y_{\sigma(n)}\}_{n=1}^\infty << \{t_{\pi(n)}\}_{n=1}^\infty.$$

It follows that

$$\{t_n\}_{n=1}^{\infty} << \{t_{\pi(n)}\}_{n=1}^{\infty},$$

and hence by Corollary VIII.b.4,

$$\{t_n\}_{n=1}^{\infty} \text{ is equivalent to } \{t_{\pi(n)}\}_{n=1}^{\infty}.$$

Thus $\{t_n\}_{n=1}^{\infty}$ is equivalent to $\{y_{\sigma(n)}\}_{n=1}^{\infty}$. □

Notes and Remarks:

1/. All of the results of this chapter carry over to T^* by the usual duality arguments. For example: every normalized unconditional basis for T^* has a permutation which dominates $\{t_n^*\}_{n=1}^{\infty}$ and has a subsequence which is permutatively equivalent to $\{t_n^*\}_{n=1}^{\infty}$.

2/. It's easily seen that the results of this chapter also hold for subsequences of $\{t_n\}_{n=1}^{\infty}$ and of $\{t_n^*\}_{n=1}^{\infty}$.

3/. Note that Corollary IX.8 is a generalization of Corollary VII.b.3.

4/. Corollary VIII.b.4 shows that T will have a U.T.A.P. normalized unconditional basis if every normalized unconditional basis for a complemented subspace of T is permutatively equivalent to a subsequence of $\{t_n\}_{n=1}^{\infty}$.

5/. Although our previous results about unconditional bases for complemented subspaces of T yield promising clues that the hypothesis in the above remark (4) might actually hold, the following proposition indicates just how delicate a proof of this might be:

Proposition IX.9:

a). If $\{x_n\}_{n=1}^{\infty}$ is any normalized unconditional basic sequence in T, then

$$\{x_n\}_{n=1}^{\infty} << \{t_{m_n}\}_{n=1}^{\infty},$$

for some $m_1 < m_2 < \cdots$.

b). If $\{x_n\}_{n=1}^{\infty}$ is a normalized unconditional basis for a subspace of T, and if T is isomorphic to a complemented subspace of $[x_n]_{n=1}^{\infty}$, then $\exists$ a permutation σ of $\mathbb{N}$ such that

$$\{x_{\sigma(n)}\}_{n=1}^{\infty} << \{t_n\}_{n=1}^{\infty},$$

and $\exists$ a subsequence $\{n_i\}_{i=1}^{\infty}$ of $\mathbb{N}$ such that $\{x_{\sigma(n_i)}\}_{i=1}^{\infty}$ is equivalent to $\{t_i\}_{i=1}^{\infty}$. (However, in general $\{x_n\}_{n=1}^{\infty}$ need not be permutatively equivalent to $\{t_n\}_{n=1}^{\infty}$.)

Proof:

a). By Theorem III.7, $[x_n]_{n=1}^\infty$ has a blocking of type T,

i.e., $\exists K > 0, p_0 = 0 < p_1 < p_2 < \cdots$, and $k_1 < k_2 < \cdots$ such that if

$$E_n := [x_i : p_{n-1} < i \le p_n],$$

and $x = \sum_n y_n \in [x_i]_{i=1}^\infty$, where $y_n \in E_n$, $(n = 1, 2, \cdots)$, then

$$\frac{1}{K} \|\sum_n \|y_n\| t_{k_n}\|_T \le \|x\| \le K \|\sum_n \|y_n\| t_{k_n}\|_T.$$

Now select natural numbers $m_1 < m_2 < \cdots$ so that

$$\begin{cases} k_n \le m_n, \text{ and} \\ \overline{\overline{E_n}} \le m_n, \\ \text{and } m_n + \overline{\overline{E_n}} \le m_{n+1}, \quad (n = 1, 2, \cdots). \end{cases}$$

Then if $x = \sum_n y_n$ (as above), where

$$y_n := \sum \{a_i x_i : p_n - 1 < i \le p_n\}, \text{ say,}$$

then

$$\|y_n\| \le \sum \{|a_i| : p_n - 1 < i \le p_n\}$$
$$\le 2\| \sum \{a_i t_i : m_n \le i \le m_n + \overline{\overline{E_n}}\} \|, \text{ for all } n = 1, 2, \cdots.$$

Thus by Corollary II.5,

$$\|x\| = \|\sum_i a_i x_i\| \le$$
$$K\|\sum_n \|y_n\| t_{k_n}\| \le$$
$$2K\|\sum_n \sum \{a_i t_i : m_n \le i \le m_n + \overline{\overline{E_n}}\} \|t_{m_n}\| \le$$
$$36K\|\sum_n \sum \{a_i t_i : m_n \le i \le m_n + \overline{\overline{E_n}}\} \|.$$

b). The classical Dvoretsky Theorem and Corollary II.5 imply $\exists$ in T a basic sequence $\{x_n\}_{n=1}^\infty$ with the property that there is a sequence of natural numbers $p_0 = 0 < p_1 < p_2 < \cdots$ such that $p_{n+1} - p_n = n$, and a sequence $k_1 < k_2 < \cdots$ such that for any $x = \sum_n a_n x_n$,

$$\tfrac{1}{20}\|\sum_n \sum \{|a_i|^2 : p_{n-1} < i \le p_n\}^{\frac{1}{2}} t_{k_n}\| \le \|x\| \le$$
$$20\|\sum_n \sum \{|a_i|^2 : p_{n-1} < i \le p_n\}^{\frac{1}{2}} t_{k_n}\|.$$

Clearly this basic sequence $\{x_n\}_{n=1}^\infty$ is not permutatively equivalent to $\underline{\text{any}}$ subsequence of $\{t_n\}_{n=1}^\infty$. To see that the first part of (b) is true, note that by (a) there is a permutation σ of $\mathbb{N}$ such that

$$\{x_{\sigma(n)}\}_{n=1}^\infty << \{t_{m_n}\}_{n=1}^\infty,$$

and since T embeds complementably in $[x_n]_{n=1}^\infty$, the proofs of Theorems IX.3 and IX.7 can be brought to bear in this case. $\qquad \square$

X. Variations on a Theme.

We give here some variant constructions for "Tsirelson-like" spaces, including a continuum of totally incomparable "Tsirelson-like" spaces, a symmetric version of T, a "treed" version, p-convexification of T, and the Tirilman spaces. (Other variations are mentioned in Chapter XI.)

X.A. The spaces $T_\theta, (0 < \theta < 1)$.

T has a continuum of analogs (the spaces T_θ) which are constructed as one might suspect.

Construction X.a.1:

Fixing $0 < \theta < 1$, T_θ is the completion of $\mathbb{R}^{(\mathbb{N})}$ with respect to the norm $\|\cdot\| := \lim_m \|\cdot\|_m$, where $\{\|\cdot\|_m\}_{m=0}^\infty$ is the monotone sequence of norms on $\mathbb{R}^{(\mathbb{N})}$ given by:

$$\begin{cases} \|x\|_0 = \max_n |a_n|, \quad \left(\text{for } x = \sum_n a_n t_n^\theta \right), \\ \|x\|_{m+1} = \max \left\{ \|x\|_m, \max \theta \cdot \sum_{i=1}^k \|E_i x\|_m \right\}, \quad (m \geq 0), \end{cases}$$

where the inner max is taken over all choices $k \leq E_1 < E_2 < \cdots < E_k$, (just as in T.)
(" $\{t_n^\theta\}_{n=1}^\infty$ " will denote the canonical unit vector basis of T_θ.)

It can easily be checked that all of our results for T also hold for $T_\theta, (0 < \theta < 1)$. An obvious question is whether or not different θ give rise to different spaces. In fact, they are quite different. Recall the following:

Definition X.a.2: Two infinite-dimensional Banach spaces X and Y are <u>totally incomparable</u> if neither contains an infinite-dimensional subspace which is isomorphic to a subspace of the other.

It's well known that the sequence spaces l_p, $(1 \leq p < \infty)$ are totally incomparable (for different p), and as a result the Baernstein spaces (see Chapter 0) B_p are totally incomparable for different p. The spaces T_θ behave the same way:

Theorem X.a.3: Let $0 < \theta < \varphi < 1$. Then T_θ and T_φ are totally incomparable.

Proof: For the sake of contradiction, fix $\theta < \varphi$ and assume $\exists$ an infinite-dimensional Banach space X which embeds in both T_θ and T_φ. By Proposition II.7, $\exists \{y_i\}_{i=1}^\infty$ in X which is equivalent to a subsequence $\{t_{n(i)}^\theta\}_{i=1}^\infty$ of $\{t_n^\theta\}_{n=1}^\infty$. Since $\{y_i\}_{i=1}^\infty$ is also equivalent to a basic sequence in T_φ, invoking Proposition II.7 once more, $\exists$ a subsequence $\{y_{i(j)}\}_{j=1}^\infty$ of $\{y_i\}_{i=1}^\infty$ which is equivalent to a subsequence $\{t_{m(j)}^\varphi\}_{j=1}^\infty$ of $\{t_j^\varphi\}_{j=1}^\infty$. It follows that $\exists$ a subsequence $\{t_{k(i)}^\theta\}_{i=1}^\infty$ of $\{t_i^\theta\}_{i=1}^\infty$ which is equivalent to a subsequence $\{t_{l(i)}^\varphi\}_{i=1}^\infty$ of $\{t_i^\varphi\}_{i=1}^\infty$.

Let $h(1) := 1$, and choose $h(2) > h(1)$ so that

$$l(h(2)) > k(h(1)).$$

Now choose $h(3) > h(2)$ so that

$$k(h(3)) > l(h(3)),$$

and choose $h(4) > h(3)$ so that

$$l(h(4)) > k(h(4)),$$

and continue choosing in this fashion.

We now have:

$$k(h(1)) < l(h(2)) < k(h(3)) < l(h(4)) < \cdots.$$

This (together with Proposition I.12) implies that

$$\{t^\theta_{k(h(j))}\}_{j=1}^\infty \approx \{t^\theta_{k(h(2j-1))}\}_{j=1}^\infty \approx$$
$$\{t^\varphi_{l(h(2j-1))}\}_{j=1}^\infty, \approx \{t^\varphi_{k(h(j))}\}_{j=1}^\infty.$$

In other words, $\exists n(1) < n(2) < \cdots$ such that

$$\{t^\theta_{n(i)}\}_{i=1}^\infty \approx \{t^\varphi_{n(i)}\}_{i=1}^\infty.$$

However, Proposition IV.c.8 shows that this is impossible. □

Notes and Remarks:

1/. Each result that we have for T generalizes to $T_\theta, (0 < \theta < 1)$.

2/. We can define "modified-T_θ" (just as we defined modified-T) and (as in Chapter V) prove that it is naturally isomorphic to T_θ.

3/. Theorem X.a.3 can be strengthened (with a good deal more effort) to show that T_θ is not crudely finitely representable in T_φ, for any $0 < \theta, \varphi < 1$, with $\theta \neq \varphi$.

X.B. Symmetric Tsirelson's Space

The first example of a symmetric Tsirelson-like space was given by T. Figiel and W. Johnson [26]. This space is a uniformly convex Banach space which contains no isomorphic copies of c_0 or l_p, $(1 \leq p < \infty)$ and possesses a symmetric basis. This construction shattered another structure-theoretic hope of the post-WWII school of Banach space theory. In the early 1970's, J. Lindenstrauss and L. Tzafriri [38, 39, 40] used an elegant fixed point argument to show that each Orlicz sequence space contained a subspace isomorphic to c_0 or some l_p, $(1 \leq p < \infty)$. This was the first class of spaces that seemed to require a sophisticated argument to find subspaces isomorphic to c_0 or l_p $(1 \leq p < \infty)$. It was then hoped that arguments of this type could be adapted to the class of symmetric spaces. The Figiel/Johnson example showed that this was not in the cards.

We present here an alternative construction for such a space, believing that it is a nice outgrowth of our study of T and that it provides a more natural and simple example of a symmetric version of Tsirelson's space.

There is a simple process for "symmetrizing" a given Banach space X with basis $\{x_n\}_{n=1}^\infty$. (N. Kalton first pointed out that this scheme would work.)

Definition X.b.1: Let X be a Banach space with basis $\{x_n\}_{n=1}^{\infty}$. We define a norm $\||\cdot\||$ on the finitely-supported elements $x = \sum_n a_n x_n$ of X by way of:

$$\||x\|| := \sup_{\sigma \in \Pi} \|\sum_n |a_n| x_{\sigma(n)}\|,$$

where Π is the family of all permutations of $\mathbb{N}$.

$S(X)$, the __symmetrization of X__, is the $\||\cdot\||$-completion of this space, with norm also denoted $\||\cdot\||$.

The vectors $\{x_n\}_{n=1}^{\infty}$ form a symmetric basis for $S(X)$ with symmetric basis constant 1. Unfortunately (as is easily seen) $S(T) \approx l_1$. However, the story about T^* is more interesting. We have:

Theorem X.b.2: Each subspace of $S(T^*)$ (with symmetric norm $\||\cdot\||$) has a subspace isomorphic to T^*. In particular, $S(T^*)$ is a reflexive Banach space with symmetric basis which does not contain a copy of c_0 or any l_p, $(1 \le p < \infty)$.

The theorem will follow immediately from the following lemmata (originally developed for Lorentz spaces), which are re-workings of results of Z. Altschuler, P. Casazza, and B. Lin [6], and Z. Altschuler [4].

Lemma X.b.3: If $y_n := \sum \{a_i t_i^* : p_n < i \le p_{n+1}\}, (n = 1, 2, \cdots)$, is a normalized block basic sequence in $S(T^*)$ such that

$$\lim_n \sup_{p_n < i \le p_{n+1}} |a_i| \ne 0, \text{ then } \lim_n \|\sum_{j=1}^n y_j\| = +\infty.$$

Proof: Assuming the hypotheses, there must exist an $\epsilon > 0$ and natural numbers $n(1) < n(2) < \cdots$ such that

$$\sup\left\{|a_i| : p_{n(j)} < i \le p_{n(j+1)}\right\} \ge \epsilon, \quad \forall j.$$

Since $\{t_n^*\}_{n=1}^{\infty}$ is symmetric in $S(T^*)$,

$$\{t_j^*\}_{j=1}^{\infty} << \{y_{n(j)}\}_{j=1}^{\infty}.$$

Thus $\exists K > 0$ such that

$$\||\sum_{j=1}^m t_j^*\||_{S(T^*)} \le K \||\sum_{j=1}^m y_{n(j)}\||_{S(T^*)}, \quad \forall m.$$

It follows that:

$$
\begin{aligned}
\|\sum_{j=1}^m t_j^*\|_{T^*} &\le \||\sum_{j=1}^m t_j^*\||_{S(T^*)} \\
&\le K\||\sum_{j=1}^m y_{n(j)}\||_{S(T^*)} \\
&\le K\||\sum_{j=1}^{n(m)} y_j\||_{S(T^*)}, \quad \forall m.
\end{aligned}
$$

But $\lim_m \|\sum_{j=1}^m t_j^*\|_{T^*} = +\infty$, so we must have

$$\lim_m \|\sum_{j=1}^m y_j\|_{S(T^*)} = +\infty.$$

$\square$

Recall from Chapter VIII, if $\{a_n\}_{n=1}^\infty$ is a sequence of reals converging to zero, we denote by $\{|\hat{a}_n|\}_{n=1}^\infty$ the non-increasing rearrangement of its non-zero elements. With this notation, we have:

Lemma X.b.4: $\exists K > 0$ such that for any $x = \sum_n a_n t_n^* \in S(T^*)$,

$$\|\sum_n |\hat{a}_n| t_n^*\|_{T^*} \leq \||x\|| _{S(T^*)} \leq K \|\sum_n |\hat{a}_n| t_n^*\|_{T^*}.$$

Proof: The first inequality follows from the definition of $\||\cdot\||$, while the second follows from Theorem VIII.a.8. $\square$

Lemma X.b.5: If $y_n := \sum \{a_i t_i^* : p_n < i \leq p_{n+1}\}$, $(n = 1, 2, \cdots)$, is a normalized block basic sequence in $S(T^*)$ with the property that $\inf_n \sup \{|a_i| : p_n < i \leq p_{n+1}\} = 0$, then $\{y_n\}_{n=1}^\infty$ has a subsequence equivalent to a block basic sequence in T^*.

Proof: Assuming the above hypotheses, since $S(T^*)$ is symmetric, we can pass to a subsequence of $\{y_n\}_{n=1}^\infty$, permute the coefficients of each y_n, re-index, shift these elements to the left, and assume that $\{y_n\}_{n=1}^\infty$ has the following properties:

(1). $y_n := \sum \{a_i t_i^* : p_n < i \leq p_{n+1}\}$, and $a_i \neq 0$ $(n = 1, 2, \cdots)$,

(2). $\inf \{|a_i| : p_n < i \leq p_{n+1}\}$

$$\geq 2^{n+1} \cdot \sup \{|a_i| : p_{n+1} < i \leq p_{n+2}\}, \quad (n = 1, 2, \cdots), \quad \text{and}$$

(3). $|a_i| \leq \frac{1}{2K \cdot p_n}$, $(p_n < i \leq p_{n+1})$, (where K is from the last lemma).

We will show that the block basic sequence $\{y_n\}_{n=1}^\infty$ in $S(T^*)$ is equivalent to the block basic sequence $\{y_n\}_{n=1}^\infty$ in T^*. By the definition of norm in $S(T^*)$,

$$\|\sum_n b_n y_n\|_{T^*} \leq \||\sum_n b_n y_n\||_{S(T^*)}, \text{ for all scalars } \{b_n\}_{n=1}^\infty.$$

To get some control in the other direction, let $\{b_n\}_{n=1}^\infty$ be a sequence of scalars such that

$$\||\sum_n b_n y_n\||_{S(T^*)} = 1.$$

Define $I := \{n : \inf \{|b_n a_i| : p_n < i \leq p_{n+1}\} \leq \inf \{|b_m a_i| : p_m < i \leq p_{m+1}\}, \text{ for some } m > n\}$. and let $J := \mathbb{N} \sim I$.

Note that for $n \in I$, (2) implies: $|b_n| \leq \frac{1}{2^{n+1}}$.

98

It follows that: $1 = |||\sum_n b_n y_n|||_{S(T^*)}$

$$\leq |||\sum \{b_n y_n : n \in I\}|||_{S(T^*)} + |||\sum \{b_n y_n : n \in J\}|||_{S(T^*)}$$

$$\leq \sum \{|b_n| \cdot |||y_n|||_{S(T^*)} : n \in I\} + |||\sum \{b_n y_n : n \in J\}|||_{S(T^*)}$$

$$\leq \tfrac{1}{2} + |||\sum \{b_n y_n : n \in J\}|||_{S(T^*)}.$$

Thus:

(4). $|||\sum_n b_n y_n|||_{S(T^*)} \leq 2|||\sum \{b_n y_n : n \in J\}|||_{S(T^*)}.$

Now order J naturally as $J := \{n(1) < n(2) < \cdots\}$. By Lemma X.b.4, and (1) we obtain:

(5). $|||\sum \{b_n y_n : n \in J\}|||_{S(T^*)} \leq K\|\sum_n b_n w_n\|_{T^*},$

where

(6). $w_n := \sum \{a_i t_i^* : q_n < i \leq q_{n+1}\}$, and

$q_1 := 0 < p_{n(1)} + 1 - p_{n(1)} =: q_2 < p_{n(1)+1} - p_{n(1)} + p_{n(2)+1} - p_{n(2)} =: q_3 < \cdots.$

By assumption (3) and the fact that $|||y_n|||_{S(T^*)} = 1, \forall n$, we have that:

$$0 = q_1 < p_1 < q_2 < p_2 \cdots. \qquad (*)$$

Also (by (3)), for each j,

$$1 = |||y_{n(j)}|||_{S(T^*)}$$

$$\leq K\|\sum \{a_{i+p_{n(j)-1}} t_i^* : 1 \leq i \leq p_{n(j)+1} - p_{n(j)}\}\|_{T^*}$$

$$\leq K \left(\tfrac{1}{2K} + \|\sum \{a_{i+p_{n(j)-1}} t_i^* : p_{n(j)} < i \leq p_{n(j)+1} - p_{n(j)}\}\|_{T^*}\right).$$

Thus, by Proposition I.16,

$$1 \leq 2\|\sum \{a_{i+p_{n(j)-1}} t_i^* : p_{n(j)} < i \leq p_{n(j)+1} - p_{n(j)}\}\|_{T^*}$$

$$\leq 16\|\sum \{a_i t_i^* : 2p_{n(j)} < i \leq p_{n(j)+1}\}\|_{T^*}.$$

Thus $\frac{1}{16} \leq \|y_{n(j)}\|_{T^*} \leq |||y_{n(j)}|||_{S(T^*)}, (j = 1, 2, \cdots)$. Now by (*) and Theorem III.5, it follows that the operator L defined by $Lw_n := y_n$ is an isomorphism, since

$$\{w_n\}_{n=1}^\infty \approx \{t_{q_n}^*\}_{n=1}^\infty \approx \{t_{p_n}^*\}_{n=1}^\infty \approx \{y_n\}_{n=1}^\infty.$$

(By Proposition II.4, these first and third equivalence constants are each 54, while Lemmas II.1 and II.3 give an equivalence constant of 6 for the middle "$\approx$".)

Thus by (5), $\exists K_1 > 0$ such that

$$|||\sum \{b_n y_n : n \in J\}|||_{S(T^*)} \leq K\|\sum_n b_n w_n\|_{T^*}$$

$$\leq KK_1\|\sum_n b_n y_n\|_{T^*}.$$

Hence

$$\|\sum_n b_n y_n\|_{T^*} \leq \|\|\sum_n b_n y_n\|\|_{S(T^*)}$$
$$\leq 2KK_1\|\sum_n b_n y_n\|_{T^*}.$$

$\square$

Finally we can give:

Proof (Theorem X.b.2):

Let X be a subspace of $S(T^*).\forall \epsilon > 0, X$ contains a basic sequence which is ϵ-close to a block basic sequence $\{y_n\}_{n=1}^\infty$ of the unit vector basis of $S(T^*)$. By Lemma X.b.3, $\{y_n\}_{n=1}^\infty$ has a block basic sequence satisfying the hypotheses of Lemma X.b.5. Thus $[y_n]_{n=1}^\infty$ (and therefore X) contains a subspace which is isomorphic to a subspace of T^*. Theorem VI.a.1 now completes the proof. $\square$

The dual of $S(T^*)$ is not explicitly known at this writing; though we would hope that the decreasing rearrangement operator D norms (up to a constant) this dual. Lemma X.b.4 shows that D norms $S(T^*)$ (up to a constant). However D does not norm $[S(T^*)]^*$, as demonstrated by:

Proposition X.b.6:

(a). For each $x \in [S(T^*)]^*, \|x\|_{[S(T^*)]^*} \leq \|Dx\|_T$.

(b). There exists no constant $K > 0$ such that for every $x \in S(T), \|Dx\|_T \leq K\|x\|_{[S(T^*)]^*}$.

Proof:

(a). If $x \in [S(T^*)]^*, \exists x^* \in S(T^*)$ so that $\|x^*\|_{S(T^*)} = 1$, and $x^*(x) = \|x\|_{[S(T^*)]^*}, (x \neq 0)$. Let $x^* = \sum_n a_n t_n^*$, and $x = \sum_n b_n t_n$. Then

$$\|x\|_{[S(T^*)]^*} = x^*(x) = \sum_n a_n b_n \leq \sum_n |\hat{a}_n| \cdot |\hat{b}_n|$$
$$\leq \|\sum_n |\hat{a}_n| t_n^*\|_{T^*} \cdot \|\sum_n |\hat{b}_n| t_n\|_T$$
$$\leq \|x^*\|_{S(T^*)} \cdot \|Dx\|_T$$
$$= \|Dx\|_T.$$

(b). For each n, let $x_n := \sum_{i=2n}^{4n-1} \frac{1}{2n} \cdot t_i$.

Then $Dx_n = \frac{1}{2n}\sum_{i=1}^{2n} t_i$, whence

$$\|Dx_n\|_T \geq \frac{1}{2n} \sum_{i=n+1}^{2n} 1 = \frac{1}{2}.$$

Fixing $K > 1$, consider $x_n \in [S(T^*)]^*$, and assume $\|x_n\|_{[S(T^*)]^*} \geq \frac{1}{K}$. Choose $x_n^* = \sum_{i=2n}^{4n-1} a_i t_i \in S(T^*)$ such that

$$\|x_n^*\|_{S(T^*)} = 1, \text{ and } x_n^*(x_n) = \|x_n\|_{S(T)}, (n = 1, 2, \cdots).$$

100

Then $\|x_n\|_{[S(T^*)]^*} = x_n^*(x_n) = \frac{1}{2n}\sum\limits_{i=2n}^{4n-1} a_i \geq \frac{1}{K}$, i.e., $\sum\limits_{i=2n}^{4n-1} a_i \geq \frac{2n}{K}$.

Thus, $\exists$ natural numbers

$$2n \leq k(1) < k(2) < \cdots < k(m) \leq 4n - 1,$$

where $m := [[\frac{2n}{2K-1}]]$, such that

$$a_{k(i)} \geq \frac{1}{2K}\|\sum\limits_{i=1}^{m} t_i^*\|_{T^*}, (i = 1, 2, \cdots, m).$$

It follows that

$$1 = \|x_n^*\|_{S(T^*)} \geq \frac{1}{2k}\|\sum\limits_{i=1}^{m} t_i^*\|_{T^*}.$$

So given $K > 0$, if we choose n so that for

$$m := [[\frac{2n}{2K - 1}]]$$

we have $\|\sum\limits_{i=1}^{m} t_i^*\|_{T^*} > 2K$, we get a contradiction. It then follows that for each $K > 1, \exists n \in \mathbb{N}$ such that

$$\|x_n\|_{S(T^*)} < \frac{1}{K}.$$

Since $\frac{1}{2} \leq \|Dx_n\|_T \leq 1, (n = 1, 2, \cdots)$, (b) follows. □

The following property of $S(T^*)$ is (as far as we know) shared by only c_0, the l_p spaces, and a special class of Lorentz spaces [19].

Proposition X.b.7: Every normalized sequence of disjointly supported vectors in $S(T^*)$ is dominated by the unit vector basis of $S(T^*)$.

Proof: Let $y_n := \sum\limits_{i \in I_n} \alpha_i t_i^*, (n = 1, 2, \cdots)$, be a disjointly supported normalized sequence in $S(T^*)$.
For any sequence of scalars $\{a_n\}_{n=1}^{\infty}$, we consider the non-increasing rearrangement

$$\{|a_n\hat{\alpha}_i| : n = 1, 2, \cdots; i \in I_n\} \text{ of}$$
$$\{a_n\alpha_i : n = 1, 2, \cdots; i \in I_n\}.$$

Let $\{J_k\}_{k=1}^{\infty}$ be a partition of $\mathbb{N}$ so that for each $k, \exists$ a permutation σ_k of J_k such that

$$\sum\limits_{n}\sum \{|a_n\hat{\alpha}_i|t_i^* : i \in I_n\} =$$
$$\sum\limits_{k} a_k \sum \{\alpha_{\sigma_k(i)}t_i^* : i \in J_k\}.$$

Now permute the J_k's in this sum so that

$$\min J_k < \min J_{k+1}, (k = 1, 2, \cdots),$$

and in this fashion obtain a permutation π and permutations π_k such that

$$\sum\limits_{k} a_k \sum \{\alpha_{\sigma_k(i)}t_i^* : i \in J_k\} =$$
$$\sum\limits_{k} a_{\pi(k)} \sum \{\alpha_{\pi_k(i)}t_i^* : i \in J_k\}.$$

Then by Lemma X.b.4,

$$\|\sum_n a_n y_n\|_{S(T^*)} \leq K\|\sum_n \sum \{|a_n \hat{a}_i| t_i^* : i \in I_n\}\|_{T^*}$$

$$\leq K\|\sum_k a_{\pi(k)} \sum \{\alpha_{\pi_k(i)} t_i^* : i \in J_k\}\|_{T^*}.$$

But if we let

$$\hat{y}_n := \sum \{\alpha_{\pi_k(i)} t_i^* : i \in J_n\},$$

then $\|\hat{y}_n\|_{T^*} \leq \|\hat{y}_n\|_{S(T^*)} = \|y_n\|_{S(T^*)} = 1, (n = 1, 2, \cdots)$. Also, since $\min J_1 < \min J_2 < \cdots$, we have

$$\min \text{ supp } \hat{y}_n \geq n.$$

Thus by Proposition V.7, $\exists K_0 > 0$ (and independent of $\hat{y}_n$) such that

$$\|\sum_k a_{\pi(k)} \sum \{\alpha_{\pi_k(i)} t_i^* : i \in J_k\}\|_{T^*}$$

$$\leq K_0 \|\sum_k a_{\pi(k)} t_k^*\|_{T^*}.$$

So finally, $\|\sum_n a_n y_n\| \leq K K_0 \|\sum_k a_{\pi(k)} t_k^*\|_{T^*}$

$$\leq \|D\| \cdot K \cdot K_0 \|\sum_k |\hat{a}_k| t_k^*\|_T$$

$$\leq \|D\| \cdot K \cdot K_0 \|\sum_k a_k t_k^*\|_{S(T^*)},$$

where D is the non-increasing rearrangement operator. $\quad\square$.

If a Banach space has a symmetric basis, then the "information" contained in the coordinates of a fixed unit vector can be used to generate a block basic sequence. More precisely

Definition X.b.8: Let X be a Banach space with symmetric basis $\{x_n\}_{n=1}^{\infty}$, and let $\alpha = \sum_n a_n x_n$ be a fixed unit vector in X. Partition the natural numbers $\mathbb{N}$ into a sequence $\{N_i\}_{i=1}^{\infty}$ of countably infinite subsets, say: $N_i := \{n_{i,1} < n_{i,2} < \cdots\}$.

Then the sequence $\{u_i^{\alpha}\}_{i=1}^{\infty}$ defined by

$$u_i^{\alpha} := \sum_j a_j x_{n_{i,j}}, (i = 1, 2, \cdots),$$

is called a <u>basic sequence generated by the vector α</u>.

It is clear that $\{u_i^{\alpha}\}_{i=1}^{\infty} >> \{x_i\}_{i=1}^{\infty}$. This fact (together with the last Proposition) yields:

Corollary X.b.9: Every basic sequence in $S(T^*)$ generated by a fixed unit vector α is equivalent to the unit vector basis of $S(T^*)$.

In fact, we can say much more:

Theorem X.b.10: The space $S(T^*)$ has a unique symmetric basic sequence.

Proof: Let $\{x_n\}_{n=1}^{\infty}$ be a normalized symmetric basic sequence in $S(T^*)$. By passing to a subsequence of $\{x_n\}_{n=1}^{\infty}$, we may assume that $\{x_n\}_{n=1}^{\infty}$ is a normalized disjointly supported sequence in $S(T^*)$ (up to an arbitrarily small perturbation). Choose sets

$$N_i := \{n_{i,1} < n_{i,2} < \cdots\}, (i = 1, 2, \cdots),$$

which partition $\mathbb{N}$ and choose scalars $a_{i,j}$ such that

$$x_i = \sum_j a_{i,j} t^*_{n_{i,j}} \in S(T^*).$$

If $\inf_i \sup \{|a_{i,j}| : j \in N_i\} = 0$, then Lemma X.b.5 implies that $\{x_i\}_{i=1}^\infty$ has a subsequence equivalent to a block basic sequence in T^*. But this is impossible, since T^* contains no symmetric basic sequences. Thus $\inf_i \sup \{|a_{i,j}| : j \in N_i\} > 0$, whence

$$\{x_n\}_{n=1}^\infty >> \{t^*_n\}_{n=1}^\infty \text{ in } S(T^*).$$

But by Proposition X.b.7,

$$\{t^*_n\}_{n=1}^\infty >> \{x_n\}_{n=1}^\infty.$$

Thus $\{x_n\}_{n=1}^\infty \approx \{t^*_n\}_{n=1}^\infty$. $\quad \square$

Notes and Remarks:

1/. Many of the results in this section have dual versions for $S(T)$. In particular, each normalized disjointly supported basic sequence in $S(T)$ dominates the unit vector basis of $S(T)$. However, a result of Z. Altshuler [5] yields that $S(T)$ does not satisfy Theorem X.b.10. Indeed $S(T)$ has a unique symmetric basis, but does not have a unique symmetric basic sequence.

2/. Z. Alshuler [4] is responsible for Definition X.b.8 and first constructed a symmetric version of T which enjoys the properties of our example.

3/. Z. Altshuler [5] has also shown that if X is a Banach space such that both it and its dual have a unique symmetric basic sequence, then X is isomorphic to c_0 or some l_p, $(1 \le p < \infty)$.

4/. We can construct "symmetric modified" Tsirelson's space, "symmetric T_θ", $(0 < \theta < 1)$, etc., and obtain similar properties for these.

5/. We can even define subsymmetric Tsirelson's space, $Su(T^*)$, as follows:

Construction X.b.11: Using the canonical unit vector basis $\{t^*_n\}_{n=1}^\infty$, let

$$Su(T^*) = \left\{ x = \sum_n a_n t^*_n : \sup \| \sum_i a_{n_i} t^*_i \|_{T^*} < \infty, \right.$$

where the "sup" is over all $n_1 < n_2 < \cdots \}$.
We norm $Su(T^*)$ by:

$$|||x|||_{Su(T^*)} := \sup \| \sum_i a_{n_i} t^*_i \|_{T^*},$$

with the "sup" as above.

It's easily seen that $\{t_n^*\}_{n=1}^{\infty}$ is a subsymmetric basis for $Su(T^*)$, and that $Su(T^*)$ has no symmetric basic sequences (and thus lacks c_0 and the spaces l_p ($1 \leq p < \infty$)).

6/. Proposition X.b.6 shows that D does not norm $[S(T^*)]^*$. It can be shown that T (with the topology generated by D) is a non-locally convex topological vector space.

X.C. "Tree-like" Tsirelson's space.

J. Lindenstrauss [33] observed that any Banach space X which contains isometric copies of $(X \oplus X \oplus \cdots \oplus X)_{l_1^n}$, ($n = 1, 2, \cdots$), must contain an isometric copy of l_1. This result yields a simple proof that there is no separable reflexive Banach space which is isometrically universal for all separable reflexive Banach spaces. Lindenstrauss then asked if, for a Banach space X, containing subspaces uniformly isomorphic to $(X \oplus X \oplus \cdots \oplus X)_{l_1^n}$, ($n = 1, 2, \cdots$), would force X to contain a copy of l_1? G. Schechtman [51] showed that this need not always occur by constructing a "tree-like" version of Tsirelson's space. His idea was to use R. C. James' method for constructing Banach spaces on trees, but to recursively define a norm as per the Figiel/Johnson construction presented in Chapter I. His construction follows:

Construction X.c.1:

a) Fix $\lambda > 1$ and let $(E, \leq)$ consist of the set

$$E := \{(n, i) : n = 0, 1, \cdots; i = 1, 2, \cdots, 2^n\},$$

with partial ordering "$\leq$" defined by:

$$(n, i) \leq (m, j) \text{ iff } n \leq m, \text{ and } (i - 1)2^{m-n} < j \leq i2^{m-n}.$$

b) Let M_0 be the set of finitely supported real functions on E.

c) For $n = 0, 1, \cdots$, and $i = 1, 2, \cdots, 2^n$, define unit vectors $e_{n,i}$ in M_0 by

$$e_{n,i}(m, j) = \begin{cases} 1, & \text{if } (n, i) = (m, j) \\ 0, & \text{otherwise.} \end{cases}$$

d) Define natural projections $P_{n,i}$ and P_n and operators $S_{n,i}$ from M_0 to M_0 by:

$$(P_{n,i}x)(m, j) = \begin{cases} x(m, j), & \text{if } (n, i) \leq (m, j), \ (x \in M_0), \\ 0, & \text{otherwise.} \end{cases}$$

$$(S_{n,i}x)(m, j) = x(m + n, (i - 1)2^m + j), \text{ and}$$

$$P_n(x) = \sum_{i=1}^{2^n} P_{n,i}(x), \ (x \in M_0).$$

e) Define now a sequence of norms $\{\|\cdot\|_m\}_{m=0}^{\infty}$ on M_0 by

$$\begin{cases} \|x\|_0 = \|x\|_{l_1} = \sum_{n,i} |x(n, i)|, \text{ and for } m \geq 0, \\ \|x\|_{m+1} = \inf\left\{\|x_0\|_m + \lambda \sum_{j=1}^{k} \max_{1 \leq i \leq 2} \|P_{j,i}x_j\|_m\right\}, \end{cases}$$

where the "inf" is taken over all $x_0, x_1, \cdots, x_k \in M_0$ for which

$$\begin{cases} \sum_{j=0}^{k} x_j = x, \text{ and} \\ P_j x_j = x_j, \text{ for all } j = 1, 2, \cdots, k, \text{ and all } k = 1, 2, \cdots. \end{cases}$$

(An induction shows: $\|x\|_{m+1} \geq \|x\|_m \geq \|x\|_0, \forall x \in M_0, \forall m \in \mathbb{N}$.)

f) $\|x\|$ can be defined as $\lim_m \|x\|_m, (x \in M_0)$. Let X be the $\|\cdot\|$-completion of M_0.

g) Now define another norm $\|\|\cdot\|\|$ on M_0 by:

$$\|\|x\|\| := \|\|x\|^2\|^{\frac{1}{2}}, \ (x \in M_0).$$

Let Y be the completion of M_0 with respect to $\|\|\cdot\|\|$. Y^* is the desired "treed" version of T.

We won't prove here that Schectman's examples do what he has claimed of them. Instead, we investigate the following question: Do we need to "tree" Tsirelson's space to get a counter-example to Lindenstrauss's question? Our next proposition shows that we must.

Proposition X.c.2:

(1) c_0 is not crudely finitely representable in T.

(2) For any $1 \leq p < \infty$, the spaces

$$(T \oplus T \oplus \cdots \oplus T)_{l_p^n}, (n = 1, 2, \cdots),$$

are not crudely finitely representable in T.

Proof:

(1) follows immediately from Lemma VII.a.3 and the fact that subspaces of a Banach space X which are uniformly isomorphic to l_∞^n, (for $n = 1, 2, \cdots$), are uniformly complemented in X.

For (2), assume that $(T \oplus T \oplus \cdots \oplus T)_{l_p^n}$ is crudely finitely representable in T with constant $C > 0$, for all $n = 1, 2, \cdots$. Let $\epsilon := \min\left\{\frac{1}{3C}, \frac{1}{2}\right\}$, and choose (by Lemma VII.a.2) $m \in \mathbb{N}$ such that every operator

$$L : [t_i]_{i=1}^m \to [t_i]_{i>m} \text{ satisfies}$$
$$\|L\| \cdot \|L^{-1}\| \geq \frac{2C}{\epsilon}.$$

Now select $n \in \mathbb{N}$ such that whenever $x_1, x_2, \cdots, x_n \in [t_i]_{i=1}^m$ and $\|x_i\| \geq \frac{1}{2}, (i = 1, 2, \cdots, n)$, then $\exists$ natural numbers $1 \leq i(1), i(2) \leq n$, with $i(1) \neq i(2)$ and $\|x_{i(1)} - x_{i(2)}\| < \epsilon$. By assumption, $\exists$ an isomorphism

$$L_1 : ([t_i]_{i=1}^m \oplus [t_i]_{i=1}^m \oplus \cdots \oplus [t_i]_{i=1}^m)_{l_p^n} \to T$$

for which $\|L_1\| \cdot \|L_1^{-1}\| \leq C$.

For each $1 \leq j \leq n$, let

$$X_j := (0, 0, \cdots, 0, [t_i]_{i=1}^m, 0, \cdots, 0),$$

105

where $[t_i]_{i=1}^m$ occurs in the j-th (of n) coordinate.

Claim: $\|Q_{m+1}|_{L_1(X_j)}\| > \epsilon$, for some $1 \leq j \leq n$, where $Q_{m+1}\left(\sum_i a_i t_i\right) := \sum_{i=m+1}^\infty a_i t_i$.

This claim must hold, since otherwise for each $1 \leq j \leq n$ there would be an $x_j \in X_j$ such that

$$\|L_1 x_j\| \leq \epsilon.$$

By our choice of n, $\exists$ natural numbers $1 \leq i(1), i(2) \leq n$ with $i(1) \neq i(2)$ and $\|P_m x_{i(1)} - P_m x_{i(2)}\| < \epsilon$, where P_m is the natural projection defined by

$$P_m\left(\sum_i a_i t_i\right) := \sum_{i=1}^m a_i t_i.$$

But then

$$\frac{1}{C} \leq \|L_1 x_{i(1)} - L_1 x_{i(2)}\|$$
$$\leq \|P_m L_1 x_{i(1)} - P_m L_1 x_{i(2)}\| + \|Q_{m+1} L_1 x_{i(1)} - Q_{m+1} L_1 x_{i(2)}\|$$
$$\leq \epsilon + 2\epsilon = 3\epsilon \leq \frac{1}{C},$$

which is a contradiction. Hence, there must be some $1 \leq j \leq n$ for which our claim holds.

Thus $X_j = [t_i]_{i=1}^m$ is $\frac{C}{\epsilon}$-isomorphic to a subspace of $[t_i]_{i>m}$. But this is impossible, by our choice of m. $\square$

T^* behaves differently from T with respect to these embeddings. Since l_∞^n embeds uniformly into T^*, every Banach space is crudely finitely representable in T^*, so this question is completely answered. Our final proposition shows what embeddings do exist into T^*.

Proposition X.c.3:

(1) For any $1 \leq p < \infty$, the spaces $(T^* \oplus T^* \oplus \cdots \oplus T^*)_{l_p^n}$ do not embed uniformly into T^*, for all $n = 1, 2, \cdots$.

(2) The spaces $(T^* \oplus T^* \oplus \cdots \oplus T^*)_{l_\infty^n}$ do not embed uniformly complementably into T^*, for all $n = 1, 2, \cdots$.

(3) The spaces $(T^* \oplus T^* \oplus \cdots \oplus T^*)_{l_\infty^n}$ embed uniformly into T^*, for all $n = 1, 2, \cdots$.

Proof:

(1). If L is an embedding of $(T^* \oplus T^* \oplus \cdots \oplus T^*)_{l_p^n}$ into T^*, then by standard perturbation arguments $\exists x_i \in (0, 0, \cdots, 0, T^*, 0, \cdots, 0)_{l_p^n}$, (where T^* is in the i-th coordinate), such that $\|x_i\| = 1, (i = 1, \cdots, n)$, and $\{Lx_i\}_{i=1}^n$ is $2\|L\|$-equivalent to a block basic sequence in $[t_i^*]_{i>n}$. Therefore

$$\|\sum_{i=1}^n Lx_i\| \leq 4\|L\|. \quad \text{On the other hand,}$$

$$\|\sum_{i=1}^n x_i\| \geq n^{1/p}.$$

Thus, $n^{1/p} \leq 4\|L\| \cdot \|L^{-1}\|$, and we obtain (1).

(2). If $(T^* \oplus T^* \oplus \cdots \oplus T^*)_{l_\infty^m}$ embeds uniformly complementably into T^*, for all $n = 1, 2, \cdots$, then $(T \oplus T \oplus \cdots T)_{l_1^m}$ embeds uniformly into T. This is impossible by Proposition X.c.2.

(3). By the dual version of Proposition V.12, $\exists$ a partition $I_1, I_2, \cdots, I_n$ of $\{n, n+1, \cdots\}$ such that

$$\left(\sum_{i=1}^n \oplus \left[t_j^* : j \in I_i \right] \right)_{l_\infty^n}$$

is 2-isomorphic to a subspace of T^*. By Theorem VI.a.1, $\exists$ a universal constant $K > 0$ such that for each $i = 1, 2, \cdots, n$,

$$\left[t_j^* : j \in I_i \right]$$

contains a subspace Y_i which is K-isomorphic to the subspace

$$(Y_1 \oplus Y_2 \oplus \cdots \oplus Y_n)_{l_\infty^n} \text{ of } \left(\sum_{i=1}^n \oplus \left[t_j^* : j \in I_i \right] \right)_{l_\infty^n},$$

which in turn is 2-isomorphic to a subspace of T^*. $\quad\square$

Notes and Remarks:

1/. A well-known problem is whether l_p embedded in a Banach space X forces l_q to embed in X^* (where $\frac{1}{p} + \frac{1}{q} = 1$). An obvious place to look for a counterexample to this conjecture is in a non-reflexive "tree-like" Tsirelson's space. This doesn't seem to have been checked into yet.

X.D. The Tirilman Spaces: $Ti\ (r, \gamma)$.

Since their introduction, the notions of type and cotype have been objects of serious study for those interested in the isomorphic theory of Banach spaces. Variant notions of equal-norm type and equal norm cotype were used by R. C. James to describe some non-reflexive uniformly non-octahedral spaces, where he used a result of G. Pisier which claims that the concepts "type 2" and "equal-norm type 2" are equivalent. L. Tzafriri (whose Romanian surname is "Tirilman") answered the question of how close these notions are in general in [57]. His answer depends upon a Tsirielson-type construction of a class of Banach spaces perverse enough to make the following hold:

Theorem X.d.1:

a) For each $1 < p < 2, \exists$ a Banach space X of cotype 2 and with a symmetric basis which is of equal-norm type p but is not of type p.

b) For each $2 < q < \infty, \exists$ a Banach space X of type 2 and with a symmetric basis which is of equal-norm cotype q but is not of cotype q.

The reader interested in the above notions and Theorem should refer to [57]. Our intent here is to give the general construction of Tirilman spaces, list their known properties (some of which we will not prove), and make some fair guesses about other properties which they might have. We begin with the construction.

Construction X.d.2:

a) Fix $0 < \gamma < 1$, and define an increasing sequence of norms $\{\|\cdot\|_m\}_{m=0}^{\infty}$ on $\mathbb{R}^{(N)}$ by way of

$$\|a\|_0 = \sup_i |a_i|, \text{ and}$$

$$\|a\|_{m+1} = \max\left\{ \|a\|_m, \sup \gamma \cdot \frac{\sum_{j=1}^{k} \|E_j a\|_m}{\sqrt{k}} \right\}, (m \geq 0),$$

where the inner "sup" is taken over *all* families of finite subsets of $\mathbb{N}$ such that

$$1 \leq E_1 < E_2 < \cdots < E_k, \text{ and over all } k \in \mathbb{N}.$$

b) An induction shows that

$$\|a\|_m \leq \|a\|_{l_2}, \forall a \in \mathbb{R}^{(N)}, \forall m \in \mathbb{N}.$$

Thus $\|a\| := \lim_m \|a\|_m$ is a norm on $\mathbb{R}^{(N)}$.

c) For a given fixed $0 < \gamma < 1$, Tirilman's space is the $\|\cdot\|$-completion of $\mathbb{R}^{(N)}$, and will be denoted "$Ti(2, \gamma)$". (For $r > 1$, replacing $k^{1/2}$ in the recursion (a) by $k^{1/r'}$, where $\frac{1}{r} + \frac{1}{r'} = 1$, we produce a space which we will denote "$Ti(r, \gamma)$").

(We let $\{x_n\}_{n=1}^{\infty}$ denote the usual unit vector basis.)

Some elementary facts concerning $Ti(r, \gamma)$ are collected in:

Proposition X.d.3: Let $r > 1 > \gamma > 0$ both be fixed. In $Ti(r, \gamma)$ we have:

(1) $\|a\| \leq \|a\|_{l_r}, \forall a \in \mathbb{R}^{(N)}$, (in fact, $\forall a \in l_r$).

(2) $\{x_n\}_{n=1}^{\infty}$ forms a 1-subsymmetric 1-unconditional basis for $Ti(r, \gamma)$.

(3) If $\nu_1, \nu_2, \cdots, \nu_k$ are pairwise disjoint normalized blocks of the $\{x_n\}_{n=1}^{\infty}$, then $\|\sum_{i=1}^{k} \nu_i\| \geq \gamma \cdot k^{1/r}$.

Proof: (We demonstrate only (1).)

Clearly, for $a \in \mathbb{R}^{(N)}, \|a\|_0 \leq \|a\|_{l_2}$.

If $\|a\|_{m-1} \leq \|a\|_{l_2}$, then for any $1 \leq E_1 < E_2 < \cdots < E_k$, and any $k \in \mathbb{N}$,

$$\frac{\gamma \sum_{j=1}^{k} \|E_j a\|_{m-1}}{\sqrt{k}} \leq \frac{\gamma \sum_{j=1}^{k} \|E_j a\|_{l_2}}{\sqrt{k}}$$

$$\leq \gamma \left(\sum_{j=1}^{k} \|E_j a\|_{l_2}^2 \right)^{1/2} \leq \gamma \cdot \|a\|_{l_2} \leq \|a\|_{l_2}.$$

Taking "sups", we have $\|a\|_m \leq \|a\|_{l_2}, \forall m$. Thus. $\|a\| \leq \|a\|_{l_2}$. In the proof nothing is sacred about $r = 2$, so we're done. $\quad\square$.

The following two lemmas of L. Tzafriri are included so that we can show that neither c_0 nor any l_p embeds in $Ti(2,\gamma)$. We include the proof of the first to indicate that the set-combinatoric tricks from Chapter I can be adapted to $Ti(2,\gamma)$. (We omit the proof of the second.)

Lemma X.d.4: If $0 < \gamma < \frac{1}{\sqrt{3}}$ and $u_1, u_2, \cdots, u_n$ are block vectors in $Ti(2,\gamma)$ with consecutive supports, then

$$\left\|\sum_{j=1}^{n} u_j\right\| \leq \sqrt{3}\left(\sum_{j=1}^{n}\|u_j\|^2\right)^{1/2}.$$

Proof: It suffices to show

(1). $\quad \left\|\sum_{j=1}^{n} u_j\right\|_m \leq \sqrt{3}\left(\sum_{j=1}^{n}\|u_j\|^2\right)^{1/2}, \quad (m = 0, 1, 2, \cdots).$

For $m = 0$, (1) is evident, so suppose that we have (1) for some m, and that $1 \leq E_1 < E_2 < \cdots < E_k$, where the E_j are finite subsets of $\mathbb{N}$.

Subdivide each E_j into (at most) three disjoint subsets such that:

a). The first is contained entirely in the support of some u_i,

b). The second contains completely the support of several blocks u_i in consecutive order, and

c). The third is contained entirely in the support of another u_i.

Let $F_1 < F_2 < \cdots < F_{k'}$ be a re-enumeration of the above formed portions of the sets E_i, where $k' \leq 3k$.

Let $I := \frac{\gamma}{\sqrt{k}}\sum_{i=1}^{k}\|E_i\left(\sum_{j=1}^{n} u_j\right)\|_m$, and note that

$$I \leq \frac{\gamma}{\sqrt{k}}\sum_{i=1}^{k'}\|F_i\left(\sum_{j=1}^{n} u_j\right)\|_m.$$

Let $A := \{1 \leq i \leq k' : F_i$ completely covers the supports of some u_j's, say $u_{p_i}, u_{p_i+1}, \cdots, u_{q_i}$, where $q_i \geq p_i\}$.

For each j, let $B_j := \{1 \leq i \leq k' : F_i \subset \text{supp } u_j\}$, and let $B := \{j : u_j$ is not covered by any F_i for which $i \in A\}$.

Then:

(2).
$$\begin{aligned}
I &= \frac{\gamma}{\sqrt{k}}\left[\sum_{i\in A}\|\sum_{j=p_i}^{q_i} u_j\|_m + \sum_{j\in B}\sum_{i\in B_j}\|F_i u_j\|_m\right] \\
&\leq \frac{\gamma}{\sqrt{k}}\left[\sum_{i\in A}\|\sum_{j=p_i}^{q_i} u_j\|_m + \frac{1}{\gamma}\sum_{j\in B}\|u_j\|_{m+1}\cdot\overline{\overline{B_j}}^{1/2}\right].
\end{aligned}$$

By our inductive hypothesis,

(3).
$$I \leq \frac{\gamma\sqrt{3}}{\sqrt{k}}\sum_{i\in A}\left(\sum_{j=p_i}^{q_i}\|u_j\|^2\right)^{1/2} + \frac{1}{\sqrt{k}}\sum_{j\in B}\|u_j\|_{m+1}\cdot\overline{\overline{B}}_j^{1/2}.$$

Applying the Cauchy-Schwartz inequality, we obtain:

(4).
$$I \leq \frac{1}{\sqrt{k}}\left(\sum_{i\in A}\sum_{j=p_i}^{q_i}\|u_j\|^2 + \sum_{j\in B}\|u_j\|^2\right)^{1/2}\bullet\left[3\gamma^2\overline{\overline{A}} + \sum_{j\in B}\overline{\overline{B}}_j\right]^{1/2} \leq$$
$$\frac{1}{\sqrt{k}}\left(\sum_{j=1}^{n}\|u_j\|^2\right)^{1/2}\cdot\left[\overline{\overline{A}} + \sum_{j\in B}\overline{\overline{B}}_j\right]^{1/2},$$

by choice of γ.

Note finally that

$$\overline{\overline{A}} + \sum_{j\in B}\overline{\overline{B}}_j \leq k' \leq 3k,$$

i.e.,

$$I \leq \sqrt{3}\left(\sum_{j=1}^{n}\|u_j\|^2\right)^{1/2},$$

and the induction is complete. □

Lemma X.d.5: Let X be an infinite-dimensional subspace of $Ti(2,\gamma)$, where $0 < \gamma < 10^{-6}$. Then for each k, $\exists w \in X$ such that $\|w\| = 1$, but

$$\frac{\gamma\sum_{j=1}^{k_0}\|E_jw\|}{\sqrt{k_0}} \leq 10\sqrt{\gamma},$$

for any choice of $1 \leq E_1 < E_2 < \cdots < E_{k_0}$, and $k_0 \leq k$.

The above two lemmas can be used to demonstrate the following result of L. Tzafriri:

Theorem X.d.6: For $0 < \gamma < 10^{-6}$, the space $Ti(2,\gamma)$ does not contain isomorphs of any l_p $(1 \leq p < \infty)$ or of c_0.

Proof:

By Proposition X.d.3(3) and Lemma X.d.4, if $\{u_j\}_{j=1}^{\infty}$ is any normalized basis for $Ti(2,\gamma)$ blocked upon $\{x_n\}_{n=1}^{\infty}$, then

$$\gamma\sqrt{n} \leq \|\sum_{j=1}^{n}u_j\| \leq \sqrt{3}\sqrt{n}, (n = 1, 2, \cdots).$$

So it suffices to show that l_2 does not embed in $Ti(2,\gamma)$.

For the sake of contradiction, suppose l_2 is isomorphic to X, some subspace of $Ti(2,\gamma)$. Then without loss of generality, we may assume that X contains a normalized block basic sequence $\{\nu_n\}_{n=1}^{\infty}$ such that

$$0.9\left(\sum_n|a_n|^2\right)^{1/2} \leq \|\sum_n a_n\nu_n\| \leq \sqrt{3}\left(\sum_n|a_n|^2\right)^{1/2},$$

for all scalar sequences $\{a_n\}_{n=1}^\infty$.

Choose now a unit vector w_1 blocked against $\{\nu_n\}_{n=1}^\infty$, and choose $N_1 \in \mathbb{N}$ such that

$$\frac{\gamma \sum_{j=1}^n \|E_j w_1\|}{\sqrt{n}} < 0.1,$$

for all choices $1 \le E_1 < E_2 < \cdots E_n$, and all $n > N_1$.

Using Lemma X.d.5, construct another unit vector w_2 supported beyond suppw_1 and also blocked against $\{\nu_n\}_{n=1}^\infty$ such that

$$\frac{\gamma \sum_{j=1}^n \|E_j w_2\|}{\sqrt{n}} \le 10\sqrt{\gamma}, \quad \text{for } n \le N_1.$$

Now choose $N_2 > N_1$ such that for $n > N_2$ and $1 \le E_1 < E_2 < \cdots < E_n$, we have

$$\frac{\gamma \sum_{j=1}^n \|E_j w_2\|}{\sqrt{n}} < 0.1.$$

Continuing in this fashion, we produce $w_1, w_2, \cdots, w_9$.

Fixing $1 \le E_1 < E_2 < \cdots < E_n$, note that

$$I(n) := \frac{\gamma \|\sum_{j=1}^n E_j(w_1 + \cdots + w_9)\|}{\sqrt{n}} \le \frac{\sum_{i=1}^9 \gamma \sum_{j=1}^n \|E_j w_i\|}{\sqrt{n}}.$$

If $1 \le n \le N_1$, then $\dfrac{\gamma \sum_{j=1}^n \|E_j w_i\|}{\sqrt{n}} \le 10\sqrt{\gamma}$,

(and this, for $i = 2, 3, \cdots, 9$), whence

$$I(n) \le 1 + 80\sqrt{\gamma}.$$

If $N_1 < n \le N_2$, then $\dfrac{\gamma \sum_{j=1}^n \|E_j w_i\|}{\sqrt{n}} < 0.1$, whence

$$I(n) \le 0.1 + 1 + 70\sqrt{\gamma}.$$

Thus for all n we have

$$I(n) \le 1.1 + 80\sqrt{\gamma},$$

which (by choice of γ) implies

$$\sup_n I(n) \le 1.1 + 80 \cdot 10^{-3}.$$

Hence, $\|w_1 + \cdots + w_9\| \le 1.1 + 80 \cdot 10^{-3}$.

But $\quad \|w_1 + \cdots + w_9\| \ge 0.9 \|w_1 + \cdots + w_9\|_{l_2}$

$$\geq \sqrt{9}(0.9) \cdot \min_{1 \leq i \leq 9} \|w_i\|_{l_2}$$

$$\geq \frac{3(0.9)}{\sqrt{3}} = \sqrt{3}(0.9) > 1.1 + 80 \cdot 10^{-3},$$

which of course is a contradiction.

The next proposition catalogs some facts about calculating the norm in $Ti(2,\gamma)$ and leads to some questions which we list in the "Notes and Remarks" section.

Proposition X.d.7:

(1) For any $x \in Ti(2,\gamma)$ and any $m \in \mathbb{N}$, either $\|x\|_m = \|x\|_0$, or

$$\|x\|_m = \sup \left(\frac{\gamma \sum_{i=1}^{k} \|E_i x\|_{m-1}}{\sqrt{k}} \right),$$

where the "sup" is over all $k \in \mathbb{N}$, and all choices $1 \leq E_1 < E_2 < \cdots < E_k$.

(2). If $0 < \gamma < 1$ and $x \in Ti(2,\gamma)$ such that

$$\|x\|_{m+1} = \frac{\gamma}{\sqrt{k}} \cdot \sum_{i=1}^{k} \|E_i x\|_m, \text{ then } \frac{1}{\sqrt{k}} \leq \gamma.$$

(3). If $x = \sum_n a_n x_n \in Ti(2,\gamma)$, then

either $\|x\| = \|x\|_0$, or $\|x\| \leq \gamma \left(\sum_n |a_n|^2 \right)^{1/2}$.

(4). If $\|x_i\| = 1, (1 \leq i \leq n)$, and supp $x_1 <$ supp $x_2 < \cdots <$ supp x_n, and $x := \sum_{i=1}^{n} x_i$, then $\|x\| = \gamma \sqrt{n}$, as long as $\gamma \sqrt{n} \geq 1$.

Proof:

(1). Follows as in Chapter I, Proposition I.10, with only notational changes.

(2). Follows from (1) by an easy argument by contradiction.

(3). We will show that for each m and each $x \in Ti(2,\gamma)$, either

$$\begin{cases} \|x\| = \|x\|_0 \text{ or} \\ \|x\|_m \leq \gamma \left(\sum_n |a_n|^2 \right)^{1/2} \end{cases}$$

For $m = 0$, this is clearly so.

If $\|x\|_{m+1} > \|x\|_m$, then

$$\|x\|_{m+1} = \sup \frac{\gamma}{\sqrt{k}} \sum_{i=1}^{k} \|E_i x\|_m$$

$$\leq \sup \gamma \sum_{i=1}^{k} \|E_i x\|_m$$

$$\leq \sup \gamma \left(\sum_{i=1}^{k} \|E_i x\|_m^2 \right)^{1/2}$$

$$\leq \sup \gamma \left(\sum_i \sum_{j \in E_i} |a_j|^2 \right)^{1/2}$$

("sups" taken as before). This last inequality follows from our inductive hypothesis, and the fact that

$$\|x\|_0 \leq \|x\|_1 \leq \|x\|_2 \leq \cdots \leq \|x\|_m.$$

Clearly, $\sup \gamma \left(\sum_i \sum_{j \in E_i} |a_j|^2 \right)^{1/2} \leq \gamma \left(\sum_n |a_n|^2 \right)^{1/2}.$

(4). $\|x\| \geq \frac{\gamma}{\sqrt{n}} \sum_{i=1}^{n} 1.$ □

We conclude with a result concerning dominance of bases in these spaces:

Proposition X.d.8: In $Ti(2, \gamma)$ every normalized block basis of $\{x_n\}_{n=1}^{\infty}$ dominates $\{x_n\}_{n=1}^{\infty}$.

Proof: Let $y_n := \sum \{\alpha_i x_i : p_n < i \leq p_{n+1}\}, (n = 1, 2, \cdots; p_0 := 0)$ be a normalized block basis of $\{x_n\}_{n=1}^{\infty}$. It suffices to show for $x = \sum_n a_n x_n$ (with supp x finite) that:

(*). $\|x\|_m = \|\sum_n a_n x_n\|_m \leq \|\sum_n a_n y_n\|, \quad (m = 0, 1, 2, \cdots).$

Now (*) holds trivially for $m = 0$, so assume that (*) holds for m.
Then $\exists 1 \leq E_1 < E_2 < \cdots < E_k$ such that

$$\|\sum_n a_n x_n\|_{m+1} = \frac{\gamma \sum_{j=1}^{k} \|E_j x\|_m}{\sqrt{k}}.$$

By our inductive hypothesis, this latter quantity is $\leq$

$$\frac{\gamma}{\sqrt{k}} \sum_{j=1}^{k} \| \sum_{n \in E_j} a_n y_n \| = \frac{\gamma}{\sqrt{k}} \sum_{j=1}^{k} \|F_j y\|,$$

where $y := \sum_n a_n y_n$, and

$$F_j := \bigcup_{n \in E_j} \{p_n + 1, p_n + 2, \cdots, p_{n+1}\}, \quad (j = 1, \cdots, k).$$

But by the definition of the norm,

$$\frac{\gamma}{\sqrt{k}} \sum_{j=1}^{k} \|F_j y\| \leq \|y\| = \|\sum_n a_n y_n\|.$$

□

Notes and Remarks:

113

1/. In the space $Ti(2,\gamma)$, (with $0 < \gamma < 1$ fixed), it can be shown that the following are equivalent:

a). $\{x_n\}_{n=1}^\infty$ is a symmetric basis in $Ti(2,\gamma)$.

b). The notions of "admissible" and "allowable" sequences of subsets of $\mathbb{N}$ generate equivalent norms on $Ti(2,\gamma)$.

It can also be shown that the following are equivalent:

a'). $Ti(2,\gamma)$ has a subspace with a symmetric basis.

b'). $\exists$ a normalized block basis

$$y_n := \sum \{\alpha_i x_i : p_n < i \le p_{n+1}\}, \quad (n = 1, 2, \cdots)$$

with the property that "admissible" and "allowable" sequences of subsets of $\mathbb{N}$ produce equivalent norms on $[y_n]_{n=1}^\infty$.

This is as far as we've gotten towards resolving:

Conjecture X.d.9: $Ti(2,\gamma)$ **has a symmetric basis.**

2/. We have been able to prove:

Proposition X.d.10: For $0 < \gamma < 1 < r$ fixed and for $0 < \epsilon < 1$ arbitrary, the unit vector basis of $Ti(r,\gamma)$ is dominated by the unit vector basis of $Ti(r, \gamma + \epsilon)$.
(The proof is a bit tedious.)

3/. An open question is whether or not "modified" Tirilman's space is naturally isomorphic to Tirilman's space.

4/. Another question is whether the unit vector basis of $Ti(2,\gamma)$ is dominated by that of T_γ, where T_γ is taken in the sense of section X.A.

5/. We can adjust the recursion in Construction X.d.2 to produce a new norm $||| \cdot |||$ as follows: fix $n \ge 1$, and let

$$|||x|||_{m+1} = \max \left\{ |||x|||_m, \gamma \cdot \sup \frac{\sum_{i=1}^{k+n} |||E_i x|||_m}{\sqrt{k}} \right\}.$$

Then the following turns out to be true:

Proposition X.d.11: If $\frac{1}{\sqrt{n}} < \gamma$, then $||| \cdot |||$ is not equivalent to $|| \cdot ||$ on $Ti(2,\gamma)$.

Proof: Define $\nu_s = \sum_{i=1}^{n^{2s}} x_i$, where $\{x_i\}_{i=1}^\infty$ is the canonical basis for $Ti(2,\gamma)$. Then by Proposition X.d.7(4),

$$||\nu_s|| = \gamma n^s.$$

114

But

$$
\begin{aligned}
|||\nu_s||| \;\geq\; &\gamma\, [|||\sum\{x_i : 1 \leq i \leq n^{2s-1}\}|||\\
&+|||\sum\{x_i : n^{2s-1} < i \leq 2n^{2s-1}\}|||\\
&+\cdots\\
&+|||\sum\{x_i : (n-1)n^{2s-1} < i \leq n \cdot n^{2s-1}\}|||]\\
=\;&n\gamma|||\sum\{x_i : 1 \leq i \leq n^{2s-1}\}|||\\
\geq\;&n^2\gamma^2|||\sum\{x_i : 1 \leq i \leq n^{2s-2}\}|||\\
\geq\;&\cdots \geq n^{2s}\gamma^{2s} = n^2(n\gamma^2)^s.
\end{aligned}
$$

But $n\gamma^2 > 1$, hence

$$
\frac{|||\nu_s|||}{\|\nu_s\|} \geq \frac{n^s(n\gamma^2)^s}{\gamma n^s} = \frac{(n\gamma^2)^s}{\gamma}, \text{ and}
$$

$$
\lim_{s\to\infty} \frac{(n\gamma^2)^s}{\gamma} = +\infty.
$$

$\square$

A consequence of this is that (for fixed n) the space generated by $||| \cdot |||$ has no bounded basic sequence dominated by the canonical unit vector basis of l_2. (To see this, note that by the above proof,

$$
\frac{|||\nu_s|||}{\sqrt{n^{2s}}} \to +\infty, \text{ as } s \to +\infty.)
$$

We believe that this space behaves like $Ti(r,\gamma)$, for $r < \frac{1}{2}$.

6/. We don't know whether Tsirelson's space embeds in any Tirilman's space.

7/. What can be said about the (complemented) subspaces of $Ti(r,\gamma)$? Does each $Ti(r,,\gamma)$ have an infinite-dimensional subspace which embeds in T_γ (the latter space in the sense of Chapter X.A.)?

8/. Any infinite-dimensional subspace of $Ti(2,\gamma)$ contains block bases (with coefficients tending to 0) of $Ti(2,\gamma)$ whose spreading models are all isomorphic to l_2.

9/. We've been able to prove (though we won't do it here):

Proposition X.d.12: Every normalized block basic sequence in $Ti(2,\gamma)$ is dominated by a subsequence of $\{t_n\}_{n=1}^\infty$, the unit vector basis for T.

From this we can deduce:

Proposition X.d.13: From each infinite-dimensional subspace W of $Ti(2,\gamma)$ we can select a basic sequence $\{\nu_n\}_{n=1}^\infty$ such that $\{\nu_n\}_{n=1}^\infty$ is dominated by some subsequence $\{t_{k_n}\}_{n=1}^\infty$ of $\{t_n\}_{n=1}^\infty$ in T.

10/. From Propositions X.d.3 and X.d.7, the following can be deduced:

Proposition X.d.14: In $Ti(2,\gamma)$, if $\{y_n\}_{n=1}^\infty$ is a normalized block basis (blocked upon $\{x_n\}_{n=1}^\infty$) and if $0 < \gamma < \frac{1}{\sqrt{3}}$, then a). $\gamma\sqrt{n} \le \|\sum_{i=1}^{n} y_i\| \le \gamma\sqrt{3}\sqrt{n}$, $(n = 1, 2, \cdots)$, and

b). If $\|\sum_n a_n x_n\| \ge K \left(\sum_n |a_n|^2\right)^{1/2}$, for all scalar choices $\{a_n\}_{n=1}^{\infty}$, then

$$K \left(\sum_n |a_n|^2\right)^{1/2} \le \|\sum_n a_n y_n\| \le \sqrt{3} \left(\sum_n |a_n|^2\right)^{1/2}$$

X.E. The p-convexifications of Tsirelson's space.

The construction of Chapter I yielded a Banach space which contained no isomorphs of c_0 or any l_p $(1 \le p < \infty)$ and no uniformly convex subspace. T. Figiel and W. Johnson [26] proceeded to build a space which failed to contain c_0 or any l_p but which was uniformly convex. This seemed to shatter what hope remained for finding reasonable structure-theoretic conditions which would ensure that a given Banach space contain isomorphs of c_0 or l_p.

In [29] W. Johnson went on to "2-convexify" the space of [26] to produce yet another "Tsirelson-like" space. It's this space (which we will denote "$T^{(2)}$" and call "convexified" Tsirelson's space) that is our concern in this section. We begin by outlining its construction.

Construction X.e.1:

Fix $1 < p < \infty$. The p-convexification of T (denoted "$T^{(p)}$") is the set of all $x = \sum_n a_n t_n$ such that

$$|x|^p := \sum_n |a_n|^p t_n \in T,$$

equipped with the norm

$$\|x\|_{(p)} = \||x|^p\|_{(p)} = \|\sum_n |a_n|^p t_n\|^{1/p}.$$

(For consistency of notation: "$T^{(1)}$" for "T", $\|\cdot\|_{(1)}$ for $\|\cdot\|_T$, and $\{t_n\}_{n=1}^{\infty}$ for the unit vector basis.)

It's easily seen that statements concerning disjointly supported blocks against the basis $\{t_n\}_{n=1}^{\infty}$ transfer (after suitable modification) from T to $T^{(p)}$. For instance: for $p > 1$, and any $x \in T^{(p)}$:

$$\|x\|_{(p)} = \max\left\{\|x\|_0, 2^{-1/p} \cdot \sup \left(\sum_{j=1}^{k} \|E_j x\|_{(p)}^p\right)^{1/p}\right\},$$

where the "sup" is over all choices

$$k \le E_1 < E_2 < \cdots < E_k.$$

The inequalities from the first few chapters suffer a bit in their constants. For instance, Proposition II.4 becomes:

Proposition X.e.2:

(a). Let $y_n := \sum_{i=p_n+1}^{p_{n+1}} a_i t_i, (n = 1, 2, \cdots)$, be a normalized block basic sequence against $\{t_n\}_{n=1}^{\infty}$ in $T^{(2)}$. Then for every choice of natural numbers

$$p_n < k_n \le p_{n+1}, (n = 1, 2, \cdots),$$

and every choice of scalars $\{b_n\}_{n=1}^{\infty}$, we have

$$3^{-1/p} \|\sum_n b_n t_{k_n}\|_{(p)} \le \|\sum_n b_n y_n\|_{(p)} \le 18^{1/p} \|\sum_n b_n t_{k_n}\|_{(p)}.$$

(b). Let $p \ge 1, E_1 < E_2 < \cdots \subset \mathbb{N}$. For each n, let L_n be a linear operator from $E_n T^{(p)}$ into itself. Then the operator $L : T^{(p)} \to T^{(p)}$ defined by

$$Lx = \sum_n L_n E_n x, (x \in T^{(p)}).$$

is bounded iff $\sup_n \|L_n\|_{(p)} < \infty$.

Moreover, $\|L\|_{(p)} \le 54^{1/p} \cdot \sup_n \|L_n\|_{(p)}$.

Our major result about $T^{(2)}$ appears in [15]:

Theorem X.e.3: $T^{(2)}$ has a unique normalized unconditional basis, up to equivalence and a permutation.

The proof would extend these notes to another chapter, so we will restrict our attention to a list of conjectures.

Notes and Remarks:

1/. Theorem X.e.3 is unknown for Tsirelson's space ("$T^{(1)}$"), but if it does hold there, it will follow that T also has a unique normalized unconditional basis, up to a permutation. The strongest result known in T is Theorem IX.3.

2/. We can easily define "modified" $T^{(p)}, (1 \le p < \infty)$, and quickly show that $T_M^{(p)}$ is naturally isomorophic to $T^{(p)}$. The interested reader is referred to [15].

3/. Since $T^{(2)}$ is the 2-convexification of T, it is 2-convex. Also, T is q-concave for some q (since otherwise, the space would fail to have a lower r-estimate, for all $1 < r$ and hence l_1^n would embed uniformly into $T^{(2)}$ on disjoint elements, which is clearly impossible). Thus, T is of type 2 and has an upper 2-estimate. Also, it can easily be shown that T is of cotype q, for every $q > 2$.

4/. In [63], W. B. Johnson shows that $T^{(2)}$ has a subspace X with the property that each subspace of every quotient of X has a basis. This is the first example of a non-Hilbertian space for which each subspace has a basis. Using the stronger properties of $T^{(2)}$ developed in these lecture notes we can show even more:

$$T^{(2)} \text{ has the property of Johnson's space } X.$$

5/. Johnson also posed the following:

Problem X.e.4: Does every subspace of $T^{(2)}$ have an unconditional basis?

(It is unknown at this time even whether there exists a non-Hilbertian space such that each of its subspaces has local unconditional structure.)

We also ask:

Problem X.e.5: Is each subspace of $T^{(2)}$ isomorphic to the span of a block basic sequence of $\{t_n\}_{n=1}^\infty$?

and we note that it can be shown:

Theorem X.e.6: Each normalized unconditional basic sequence in $T^{(2)}$ is permutatively equivalent to the span of a disjointly supported sequence in $T^{(2)}$.

6/. In Chapter VIII we demonstrated that the non-increasing rearrangement operator was bounded on T. Similar reasoning can be used to show that it is also bounded on $T^{(p)}, (1 < p < \infty)$, and norms $S(T^{(p)})$. Furthermore, (for such p), $\left[S\left((T^{(p)})^*\right)\right]^* \approx S(T^{(p)})$.

7/. It can easily be shown (using techniques from X.B.) that $S(T^{(p)}), (1 < p < \infty)$, is a Banach space with symmetric basis such that:

a) Each subspace of $S(T^{(p)})$ contains a subspace isomorphic to a subspace of $T^{(p)}$. Hence, $c_0, l_q, (1 \le q < \infty)$, do not embed into $T^{(p)}$.

b) $S(T^{(p)})$ is normed by the non-increasing rearrangement operator D.

c) $[S(T^{(p)})]^* \approx S[(T^{(p)})^*]$, and the former is also normed by D.

d) Lemma X.b.3 holds in $S[(T^{(p)})^*]$, while Lemma X.b.4 and X.b.5 hold in $S(T^{(p)})$ and $S[(T^{(p)})^*]$, and Corollary X.b.9 holds in $S[(T^{(p)})^*]$.

8/. It is immediate that $S(T^{(p)}), (1 < p < 2)$, is of cotype 2 and equal norm type p, but not type p. This gives an alternate example for Theorem X.d.1.

118

Chapter XI: Some final comments

There are several other places where Tsirelson's space (or a variant) has appeared, and we would be remiss not to mention these occurences.

P. G. Casazza, B. L. Lin, and R. H. Lohman [20] (simultaneously with B. Beauzamy [10]) constructed a version of T which is of co-dimension one in its bidual. This "James-Tsirelson" space lacks embedded copies of c_0 and the l_p spaces ($1 \leq p < \infty$). Summing this space with itself k times produces versions which are of co-dimension k in their second duals and yet still lack embedded isomorphs of c_0 and the l_p spaces.

J. Elton, P.K. Lin, E. Odell, and S. Szarek [25] have shown that both T and T^* have the fixed point property.

Spreading models of T, T^*, the 2-convexification of T, James-Tsirelson space, etc. have been studied extensively by B. Beauzamy and J. T. Lapreste in [12].

In [1], R. Alencar, R. Aron, and S. Dineen have shown that T^* has the property that $H(T^*)$ is reflexive. ($H(X)$ is the space of holomorphic functions on X equipped with the T_ω topology.) Thus T^* becomes the first known example of an infinite-dimensional Banach space for which the space of holomorphic functions is reflexive. They also show that $H(T)$ is not reflexive. It is curious to us that this is the only area of analysis wherein the l_p spaces are pathological, while Tsirelson's construction yields an example of a space with "good" properties.

R. Alencar, R. Aron, and G. Fricke [2] have studied tensor products of Tsirelson's space. Their primary result: the completion of the n-fold injective tensor product of T^* lacks embedded copies of the l_p spaces ($1 < p < \infty$), has no unconditional basis, and in fact even fails local unconditional structure.

B. Beauzamy [10] has spoken with us about "Lorentz-Tsirelson" space and "Orlicz-Tsirelson" space, describing the latter as the "ultimate horror".

E. V. Tokarev [61] has produced a function space which lacks embedded copies of c_0 and the l_p spaces ($l \leq p < \infty$).

The computer program in the appendix is bound to raise some eyebrows, yet anyone who has calculated the norm of a concrete vector in T by hand cannot fail to be curious about the possibility of using this program (or some refinement) to sharpen some working hypotheses into conjectures. Time will be the best judge of whether or not such devices might help advance the theory.

We believe that Tsirelson's space and its variations still have a lot of life left in them. The space of Kalton and Peck [32] ("KP") is believed by some to be a counter-example to the hyperplane conjecture: is every infinite-dimensional Banach space isomorphic to its hyperplanes? However, KP is not easy to work with because KP is hereditarily-l_2. This property of KP produces large numbers of isomorphisms between different subspaces of KP and makes checking the hyperplane problem quite difficult. Since Tsirelson's space has so few isomorphisms between its subspaces, it might be easier to check the hyperplane conjecture against a "twisted-sum" of symmetrized Tsirelson's spaces.

T. Odell [62] has just produced a "Tsirelson-type" space that is non-separable and contains no subsymmetric basic sequences.

Finally, in mathematics good conjectures often survive their resolutions as refined conjectures. The example of B.S. Tsirelson finally demonstrated that an infinite-dimensional Banach space need not contain an isomorph of c_0 or of any l_p. H. Rosenthal [49] has given a reformulation of this question (which he calls the "Problem"): *Must every infinite-dimensional Banach space contain an infinite-dimensional sub- space which is isomorphic to c_0, some l_p, or some reflexive space?*

In fact Krivine [49] has given a partial positive response to the conjecture which was countered by Tsirelson's example:

Theorem (Krivine): Let $\{x_j\}_{j=1}^{\infty}$ be a sequence in a Banach space with infinite-dimensional span. Then either: $\exists\ 1 \leq p < \infty$ such that l_p is block-finitely represented in $\{x_j\}_{j=1}^{\infty}$, or: c_0 is block-finitely represented in some permutation of $\{x_j\}_{j=1}^{\infty}$.

Bibliography: **Tsirelson's Space**

1. Alencar, R., Aron, R. and Dineen, S.: A reflexive space of holomorphic functions in infinitely many variables, Proc. A.M.S., (90), 1984, pp.407-411.

2. Alencar, R., Aron, R. and Fricke, G.: Tensor products of Tsirelson's space, (pre-print).

3. Alspach, D.: Quotients of c_0 are almost isometric to subspaces of c_0, Proc. A.M.S., (76), 1979, pp. 285-288.

4. Altshuler, Z.: A Banach space with a symmetric basis which contains no l_p or c_0, and all of its symmetric basic sequences are equivalent, Compositio Mathematica, (35), 1977, pp.189-195.

5. ————: Characterization of c_0 and ℓ_p among Banach spaces with symmetric bases, Israel Journal of Math., (24), 1976, pp.39-44.

6. Altshuler, Z., Casazza, P. G. and Lin, B.: On symmetric basic sequences in Lorentz sequence spaces, Israel Journal of math., (15), 1973, pp.140-155.

7. Aron, R., Baker, J.W., Murphy, T. and Slotterbeck, O.A.: A program for calculating the norm in Tsirelson's space, (pre-print).

8. Baernstein, A.: On reflexivity and summability, Studia Mathematica, (42), 1972, pp.91-94.

9. Banach, S. and Saks, S.: Sur la convergence forte dans les champs L^P, Studia Mathematica, (2), 1930, pp.51-57.

10. Beauzamy, B.: Deux espaces de Banach et leurs modeles etales, University of Lyon, 1980.

11. ————: Espaces d'Interpolation reels, Lecture Notes (666), Springer-Verlag, New York.

12. Beauzamy, B. and Lapreste, J.T.: Modeles etales des espaces de Banach, Publications du Department de Mathematiques, University of Lyon, 1983.

13. Bellenot, S.: The Banach space T and the fast growing hierarchy from logic, Israel Journal of Math., (47), 1984, pp.305-313.

14. ————: Tsirelson superspaces and ℓ_p, (pre-print).

15. Bourgain, J., Casazza, P. G., Lindenstrauss, J. and Tzafriri, L.: Banach spaces with a unique unconditional basis,, up to a permutation, Memoirs of the A.M.S., No. 322, 1985.

16. Casazza, P. G.: Tsirelson's space, Proc. of the workshop on Banach space theory, (1981), B.L. Lin, ed.

17. Casazza, P. G., Johnson, W. and Tzafriri, L.: On Tsirelson's space, Israel Journal of Math., (47), 1984, pp.81-98.

18. Casazza, P. G. and Lin, B.: On symmetric basic sequences in Lorentz sequence spaces II, Israel Journal of Math., (17), 1974, pp.191-218.

19. ————: Perfectly homogeneous bases in Banach spaces, Canad. Math. Bull., (18), 1975, pp.137-140.

20. Casazza, P. G., Lin, B. L. and Lohman, R. H.: On nonreflexive Banach spaces which contain no c_0 or ℓ_p, Can. Journal Math., (32), 1980, pp.1382-1389.

21. Casazza, P. G. and Odell, E.: Tsirelson's space and minimal subspaces, Longhorn Notes, University of Texas, 1982-1983.

22. Casazza, P. G.: Finite dimensional decompositions in Banach spaces, in: Geometry of Normed Linear Spaces, editors: R. G. Bartle, N. T. Peck, A. L. Peressini, and J. J. Uhl; Contemporary Mathematics, Volume 52, 1986.

23. Casazza, P. G. and Odell, E.: On Tsirelson's space II, (pre-print).

24. Edelstein, M. and Wojtaszczyk, P. G.: On projections and unconditional bases in direct sums of Banach spaces, Studia Mathematica, (56), 1976, pp.263-276.

25. Elton, J., Lin, P. G.K., Odell, E. and Szarek, S.: Remarks on the fixed point problem for non-expansive mappings, in: Fixed points and non-expansive mappings, Contemporary Math., Volume 18, (1983), pp.87-120.

26. Figiel, T. and Johnson, W. B.: A uniformly convex Banach space which contains no ℓ_p, Compositio Math., (29), 1974, pp.179-190.

27. Johnson, W. B.: A reflexive Banach space which is not sufficiently Euclidean, Studia Mathematica, (55), 1976, pp.201-205.

28. ———: On quotients of L^p which are quotients of ℓ_p, Compositio Math., (33), 1976, pp.

29. ———: Banach spaces all of whose subspaces have the approximation property, Special topics of applied math., 1980, North Holland Pub., Amsterdam, pp.15-26.

30. Johnson, W. B.and Zippin, M.: Subspaces and quotient spaces of $(\sum G_n)_{\ell_p}$ and $(\sum G_n)_{c_0}$. Israel Journal of Math., (17), 1974, pp.50-55.

31. ———: On subspaces of quotients of $(\sum G_n)_{\ell_p}$ and $(\sum G_n)_{c_0}$, Israel Journal of Math., (13), 1972, pp.311-316.

32. Kalton, N. and Peck, N.: Twisted sums of sequence spaces and the three space problem, Transactions A.M.S.,, (255), 1979, pp.1-30.

33. Lindenstrauss, J.: Notes on Klee's paper "Polyhedral sections of convex bodies', Israel Journal of Math., (4), 1966, pp.235-242.

34. Lindenstrauss, J. and Pelczynski, A.: Absolutely summing operators in L_p spaces and their applications, Studia Mathematica, (29), 1968, pp.275-326.

35. Lindenstrauss, J. and Tzafriri, L.: Classical Banach Spaces I: sequence spaces, Ergebnisse der Mathematik, #92, 1977, Springer-Verlag, New York.

36. ———: Classical Banach Spaces II: function spaces, Ergebnisse der Mathematik, #97, 1979. Springer-Verlag, New York.

37. ———: On the complemented subspaces problem, Israel Journal of Math., (9), 1971, pp.263-269.

38. ———: On Orlicz sequence spaces, Israel Journal of Math., (10), 1971, pp.379-390.

39. ———: On Orlicz sequence spaces II, Israel Journal of Math., (11), 1972, pp.355-379.

40. ———: On Orlicz sequence spaces III, Israel Journal of Math., (14), 1973, pp.368-389.

41. ———: On the isomorphic classification of injective Banach lattices, Math. Analysis and Applications, Volume 7B, 1981, Academic Press, New York, pp.489-498.

42. ———: The uniform approximation property in Orlicz spaces, Israel Journal of Math., (23), 1976, pp.142-155.

43. Maurey, B. and Pisier, G.: Series de variables aleatoires vectorielles independants et propriettes geometriques des espaces de Banach, Studia Mathematica, (58), 1976, pp.45-90.

44. Odell, E.: On the types in Tsirelson's space, Longhorn Notes, University of Texas, 1982-1983.

45. Pelczynski, A.: Projections in certain Banach spaces, Studia Mathematica, (19), 1960, pp.209-228.

46. Pelczynski, A. and Rosenthal, H.: Localization techniques in L^P spaces, Studia Mathematica, (52), 1975, pp.263-289.

47. Retherford, J. and Stegall, C.: Fully nuclear and completely nuclear operators with applications to L_1 and L_∞ spaces, Transactions A.M.S., (163), 1972, pp.457-492.

48. Rosenthal, H.: On subspaces of L^P, Annals of Math., (97), 1973, pp.344-373.

49. ———: On a Theorem of Krivine concerning block finite representability of ℓ_p in general Banach spaces, J. Func. Anal., (28), 1978, pp.197-225.

50. ———: Some recent discoveries in the isomorphic theory of Banach spaces, Bull. A.M.S., (84), 1978, pp.803-831.

51. Schechtman, G.: A tree-like Tsirelson space, Pacific Journal Math., (83), 1979, pp.523-530.

52. Schreier, J.: Ein Gegenbeispiel zur Theorie der schwachen Konvergenz, Studia Mathematica, (2), 1930, pp.58-62.

53. Seifert, C.J.: Averaging in Banach spaces, (dissertation), Kent State University, 1977.

54. ———: The Dual of Baernstein's space and the Banach-Saks Property, Bulletin de l'Academie Polonaise des Sciences, XXVI, no.3, pp.237-239, 1978.

55. Smorynski, C.: "Big" news from Archimedes to Friedman, Notices A.M.S., (30), 1983, pp.251-256.

56. Tsirelson, B.S.: Not every Banach space contains an embedding of ℓ_p or c_0, Functional Anal. Appl., (8), 1974, pp.138-141 (translated from the Russian).

57. Tzafriri, L.: On the type and cotype of Banach spaces, Israel Journal of Math., (32), 1979, pp.32-38.

58. ———: Some directions of research in Banach space theory, in Functional Analysis: Surveys and Recent Results II, ed. by K. Bierstent and B. Fuchssteiner, 1980, North Holland

Pub., Amsterdam.

59. van Dulst, D.: Reflexive and super-reflexive Banach spaces, Tract 102, Mathematisch Centrum, Amsterdam, 1982.

60. Zippin, M.: On perfectly homogeneous bases in Banach spaces, Israel Journal of Math., (4), 1966, pp.265-272.

61. Tokarev, E. V.: A symmetric Banach space of functions, not containing ℓ_p $(1 \leq p < \infty)$ and c_0, Functional Anal. Appl., No. 2, 1984, pp.150-151 (translated from the Russian).

62. Odell, E.: No subsymmetric sequences, Longhorn Notes, University of Texas, 1984-1985.

63. Johnson, W. B.: Banach spaces all of whose subspaces have the approximation property. Seminare d'Analyse Fonct, Expose 16 (1979-80), Ecole Polytechnique, Paris.

WEAK HILBERT SPACES:

An Appendix

While going to press with this manuscript, we were approached with the possibility of inserting some recent strong results about "weak Hilbert spaces". Convexified Tsirelson's space turns out to be a primal example of such an object, and so we include here a brief treatment of such spaces, claiming neither completeness nor rigor in this treatment. A separate bibliography follows; we have organized this material as follows:

> Aa: Type and Cotype.
>
> Ab: Convexified Tsirelson's Space.
>
> Ac: Weak Type and Weak Cotype.
>
> Ad: Weak Hilbert Spaces.
>
> Ae: Constructibility Properties of Weak Hilbert Spaces.
>
> Af: Open Problems.

Aa: Type and cotype

Recently G. Pisier [26] introduced the notion of a "weak Hilbert space". At this writing, convexified Tsirelson's space $T^{(2)}$ and its dual provide the main non-trivial examples of such spaces. Our hope here is to relate Pisier's general theory to what's known about $T^{(2)}$ and to some open problems. In this initial section we will introduce the Banach space notions of type and cotype and review some of the main results of their theory. We will not prove the theorems of this section, since they are readily accessible elsewhere. These ideas provide a framework for variant results in the theory of weak Hilbert spaces. Appropriate source materials can be found in [13, 18, 24, 25, 26, and 29]. We begin with type and cotype for operators on a Banach space.

Definition Aa1: Let X, Y be Banach spaces, and $U : X \to Y$ an operator.

1. U is of type p, $(1 \leq p \leq 2)$, if there is a constant $M > 0$ such that for all finite subsets $\{x_1, x_2, \ldots, x_n\}$ of X,

$$\int_0^1 \left\| \sum_{j=1}^n r_j(t) x_j \right\| dt \leq M \left(\sum_{j=1}^n \|U x_j\|^p \right)^{1/p},$$

where $\{r_j\}_{j=1}^\infty$ is the sequence of Rademacher functions.

("$T_p(U)$" is the infimum of all such M).

Furthermore, if U is the identity from X to X, and $T_p(U) < \infty$, we say:

X is of type p, and write: "$T_p(X)$", for $T_p(U)$.

2. We say that U is of cotype q, $(2 \leq 2 \leq \infty)$, if there is a constant $N > 0$ such that for all finite subsets $\{x_1, x_2, \ldots, x_n\}$, of X,

$$\frac{1}{N} \left(\sum_{j=1}^{n} \|Ux_j\|^q \right)^{1/q} \leq \int_0^1 \left\| \sum_{j=1}^{n} r_j(t)x_j \right\| dt.$$

We denote by "$C_q(U)$", the infimum of all such N. As in (1), we define what is meant by claiming X is of cotype q, and "$C_q(X)$".

It is well known that:

Theorem Aa2: X of type $p \Rightarrow X^*$ of cotype q, for $\frac{1}{p} + \frac{1}{q} = 1$.

However, ℓ_1 is of cotype 2, while neither its dual nor its predual have any type or cotype.

Our next definitions allow us to study spaces which embed into other spaces.

Definition Aa3: If X and Y are Banach spaces, the Banach-Mazur distance from X to Y (denoted "$d(X, Y)$") is

$$d(X, Y) = \inf\{\|T\| \cdot \|T^{-1}\| : T : X \to Y \text{ is a linear operator}\}.$$

If Y is a Hilbert space with $\dim(X) = \dim(Y)$, we write d_X for $d(X, Y)$. (Note that this is not a true "metric" distance, since $d(X, X) = 1$, but $\log d(X, Y)$ will produce a usable metric.)

Definition Aa4: Let X be a Banach space and $\{E_n\}_{n=1}^{\infty}$ any sequence of Banach spaces. We say X contains E_n uniformly (respectively: E_n is uniformly complemented in X) if there exists $\lambda > 0$ and sequence $\{F_n\}_{n=1}^{\infty}$ of subspaces of X such that $d(E_n, F_n) \leq \lambda, \forall n$ (respectively: there exist projections $P_n : X \to F_n$, onto, such that $\sup_{n \geq 1} \|P_n\| < \infty$).

We generally use this definition for $E_n := \ell_p^n$, for some $1 \leq p < \infty$. In particular, G. Pisier [25] has shown:

Theorem Aa5: If a Banach space X does not contain the spaces ℓ_1^n uniformly, then X (respectively, X^*) is of type p iff X^* (respectively, X) is of cotype q with $\frac{1}{p} + \frac{1}{q} = 1$.

Clearly, any Banach space which is isomorphic to a Hilbert space must be of type 2 and cotype 2. Kwapien [24] proved the converse:

Theorem Aa6: A Banach space X is isomorphic to a Hilbert space iff X is of type 2 and cotype 2. Moreover,

$$d_X \leq T_2(X)C_2(X).$$

(When we eventually introduce the notions of "weak type 2" and "weak cotype 2" we will use a "weak" version of Theorem Aa6 to define a "weak" Hilbert space.) The following result [24] clarifies the meanings of type and cotype while relating these notions to embeddings of ℓ_p^n's into the space:

Theorem Aa7:

A. Let X be an infinite dimensional Banach space,

$$\text{and } p(X) : \; = \sup\{p : X \text{ is of type } p\},$$
$$\text{and } q(X) : \; = \inf\{q : X \text{ is of cotype } q\},$$

Then X contains the spaces ℓ_p^n uniformly for both $p := p(X)$ and $p := q(X)$.

B. Maurey [15] proved an important theorem on extensions of operators which is basic to the notion of weak Hilbert spaces (which we haven't even defined yet!):

Theorem Aa8: Let X be a Banach space of type 2. Then for any subspace Y of X and any bounded operator $U : Y \to H$, (where H is some Hilbert space), there is an extension $\bar{U} : X \to H$ such that $\|\bar{U}\| \le T_2(X) \cdot \|U\|$.

This theorem motivates the following definition:

Definition Aa9: A Banach space X has the Maurey Extension Property ("M.E.P.") if there exists a constant $K > 0$ such that for every subspace S of X and every operator $U : S \to H$, where H is a Hilbert space, there exists an extension $\bar{U}$ of U such that $\bar{U} : X \to H$ and $\|\bar{U}\| \le K\|U\|$.

In this language, Theorem Aa8 states that every type 2 space has M.E.P. The converse is a famous open problem in this field:

Problem Aa10: If a Banach space X has M.E.P., must it have type 2?

We also note that if a space X has M.E.P. and Y is a subspace with $d_Y < \infty$, then there exists a Hilbert space H and an isomorphism $T : Y \to H$ which can be extended to a bounded operator $\hat{T} : X \to H$. It follows that $P := T^{-1} \circ \hat{T}$ is a projection of X onto H, i.e., all subspaces of X isomorphic to Hilbert spaces are complemented in X. Moreover, if $\{E_n\}_{n=1}^{\infty}$ is a sequence of finite dimensional subspaces of X and $\sup_n d_{E_n} < \infty$, then the E_n's are uniformly complemented in X. This leads us to define:

Definition Aa11: A Banach space X has the Maurey Projection Property ("M.P.P.") if there exists a function $f : R^+ \to R^+$ such that for any subspace Y of X with $d_Y < \infty$, there exists a projection $P : X \to Y$ with $\|P\| \le f(d_Y)$.

Our earlier discussion shows that, for a given Banach space X, M.E.P. implies M.P.P. The converse is open.

Problem Aa12: Does M.P.P. imply M.E.P.? (or, does M.P.P. for a space X imply that X is of type 2?).

We now recall an important result of Szankowski [24]. It helps explain some of the behavior of weak Hilbert spaces.

Theorem Aa13: If X is a Banach space and every subspace of X has the approximation property, then $p(X) = q(X) = 2$.

Another area of study needed to develop the notion of weak Hilbert spaces is embeddings of ℓ_2^n into a Banach space. In 1962 Dvoretsky [26] proved the following:

Theorem Aa14: If X is an infinite dimensional Banach space, then for every $\epsilon > 0$, and every $n = 1, 2, \ldots$, there is an n-dimensional subspace E_n of X so that $d_{E_n} \leq 1 + \epsilon$.

In 1977 Kashin [18] showed:

Theorem Aa15: For each n, and $\epsilon > 0$ there are two n-dimensional subspaces of ℓ_1^{2n} (say E_1, E_2) which are orthogonal in the sense of ℓ_1^{2n} and so that $d_{E_i} \leq 1 + \epsilon, (i = 1, 2)$.

It is well-known that such subspaces cannot be well complemented. In a study in 1977 T. Figiel, J. Lindenstrauss, and V. Milman [6] found the "best" embeddings of spaces ℓ_2^n into finite dimensional spaces. One of their many results was:

Theorem Aa16: There is a $c > 0$ so that for $X := \ell_p^n$, there is a subspace Y of X with dim $Y = k$ and $d_Y \leq 2$, where:

(a.) $k = c \log n$, if $p = \infty$,

(b.) $k = c n^{2/p}$, if $2 \leq p < \infty$, and

(c.) $k = cn$, if $1 \leq p \leq 2$.

Moreover, each of these is best possible.

These authors then approached the issue of embedding ℓ_2^n into spaces of cotype 2, which was rapidly pushed by several authors to the result of S. Dilworth and S.J. Szarek [5]:

Theorem Aa17: Every $2n$-dimensional normed space E contains two n-dimensional subspaces E_1 and E_2 which are orthogonal with respect to the John ellipsoid of E and which satisfy

$$d_{E_i} \leq f(C_2(E)),$$

where $f(C_2(E))$ is a number which depends only on the cotype constant of $E, (i = 1, 2)$.

That the converse of the above result fails (even in a stronger sense) was observed by W.B. Johnson, and will be discussed later.

Notes and Remarks:

1/. B. Maurey [15] observed a generalization of Theorem Aa8 which gives the following fundamental result on factorizations:

Theorem Aa18: If X is of type 2 and Y is of cotype 2, then every operator $U : X \to Y$
factors through a Hilbert space.

2/. Related to Problem Aa12, we have:

Proposition Aa19: If the space X has M.P.P., then $\sup\{p|X$ is type $p\} = 2$.

Proof: If $\sup\{p|X$ is type $p\} = p_o < 2$, then by Theorem Aa7, $\{\ell_{p_o}^n\}_{n=1}^\infty$ embed uniformly into
X (in the sense of Definition Aa4). By Theorem Aa16, there exists (for each n) a subspace
E_n of $\ell_{p_o}^n$ with $d_{E_n} \le 2$ and $\dim E_n \ge cn$, for a universal c. Now since X has M.P.P., these E_n
are uniformly complemented in $\ell_{p_o}^n$. Hence E_n^* is uniformly complemented in $(\ell_{p_o}^n)^* = \ell_{p_o'}^n$ where
$\frac{1}{p_o} + \frac{1}{p_o'} = 1$. But $d_{E_n^*} \le 2$, and $\dim E_n^* = \dim E_n \ge cn$, which contradicts (a) of Theorem Aa16.
□

Ab: Convexified Tsirelson's Space.

Recall here the construction in section X.E of $T^{(2)}$, the 2-convexification of Tsirelson's space.
Using notation from that section, for $x = \sum_n a_n t_n$ in $T^{(2)}$, it follows that

$$\|x\|_{(2)} = \max\left\{ \|x\|_o, \quad 2^{-\frac{1}{2}} \sup\left\{ \sum_{j=1}^k \|E_j x\|_{(2)}^2 \right\}^{1/2} \right\},$$

where the "sup" is over all admissible sums for x.

Proposition Xe2 and Corollary IVb2 hold for $T^{(p)}, (1 \le p < \infty)$, with only notational changes
in their proofs. This yields our next proposition (which employs the "fast growing hierarchy"
in its statement).

Proposition Ab1: If $\{y_k\}_{k=1}^{f_i(n)}$ is a normalized sequence of disjointly supported vectors in $T^{(2)}$
with support $(y_k) \subset \{n, n+1, \ldots\}$, then $\{y_k\}_{k=1}^{f_i(n)}$ is $C \cdot 2^i$-equivalent to the unit vector basis of
$\ell_2^{f_i(n)}$, where C is a universal constant. Moreover,

$$[y_k : 1 \le k \le f_i(n)] \text{ is } f(C \cdot 2^i) - \text{complemented in } T^{(2)}.$$

The constant C is the equivalence constant between $T^{(2)}$ and "modified" $T^{(2)}$. (See (2),
Notes/remarks of X.e). Propositions V.6 and Ab1 now yield:

roposition Ab2: There is a universal constant $C_o > 0$ such that whenever $E \subset [t_j]_{j=n}^\infty$ and
$\dim E \le f_i(n)$, then

(i) $d_E \le C_o 2^i$, and

(ii) there exists a projection $P : T^{(2)} \to E$, onto, such that

$$\|P\| \le C_o \cdot 2^i$$

One should note that both Propositions Ab1 and Ab2 also hold in $(T^{(2)})^*$. We are now ready to prove:

Theorem Ab3: $T^{(2)}$ is of type 2 and cotype q, for all $q > 2$.

Proof: Let $q > 2$ and assume for each n that ℓ_q^n is K-isomorphic to a subspace E_n of $T^{(2)}$. Defining $F_n := E_n \cap [t_j]_{j=[[\frac{n}{2}]]}^{\infty}$, then $\dim F_n \geq \frac{n}{2}$, and so by Proposition Ab2, $d_{F_n} \leq 2C_o$, and F_n is $2C_o$-complemented in $T^{(2)}$ (and hence in E_n). Now Theorems Aa7 and Aa16 imply that $\inf\{q > 2 | T^{(2)} \text{ is cotype } q\} = 2$. Since $T^{(2)}$ is cotype $q > 2$ and is the 2-convexification of a Banach lattice, it follows (see, e.g., [13], section 1.f) that $T^{(2)}$ is type 2. $\square$

It follows from Theorem Aa5:

Theorem Ab4: $(T^{(2)})^*$ is of cotype 2, and of type p, for every $1 \leq p < 2$.

Since ℓ_2 does not embed into $T^{(2)}$ or its dual (or even into a subspace of a quotient of either), we have that $T^{(2)}$ is not of cotype 2 and $(T^{(2)})^*$ is not of type 2, by Kwapien's Theorem Aa6. Finally, by Proposition Ab2, we obtain:

Theorem Ab5: For every $0 < \delta < 1$, there is a constant $C(\delta) > 0$ such that for any finite dimensional subspace E of $T^{(2)}$,

(i) there is a subspace F of E with $\dim F \geq \delta \cdot \dim E$, and $d_F \leq C(\delta)$,

(ii) and a projection $P : T^{(2)} \to F$, onto, with $\|P\| \leq C(\delta)$.

(Moreover, the same holds in $(T^{(2)})^*$).

Proof: Choose an i so that $f_i((1 - \delta)n) \geq n$, for all large n. Given a subspace E of $T^{(2)}$ with $\dim E = n$, let $k := [[(1 - \delta)n]]$, and define P_k by $P_k\left(\sum_j a_j t_j\right) := \sum_{j=1}^{k} a_j t_j$. Then $P_k|_E$ has a kernel of $\dim \geq \delta n$. Letting $F := \ker(P_k|_E)$, then

$$\delta n \leq \dim F \leq n, \quad F \subset [t_j]_{j=k}^{\infty}, \text{ and } f_i(k) \geq n.$$

By Proposition Ab2, the result follows. (Also, since Proposition Ab1 and Ab2 hold in $(T^{(2)})^*$, so does Theorem Ab3.) $\square$

The proof of Proposition IV.b.4 works in $T^{(2)}$ and $(T^{(2)})^*$ to show:

Theorem Ab6: For all $m = 1, 2, \ldots$, in $T^{(2)}$ (and its dual):

$$d_{[t_i]_{i=1}^n} = o(\log_m n), \quad \text{where}$$

$$\log_1 n := \log n, \text{ and } \log_{m+1}(n) := \log(\log_m n).$$

This observation, together with the proof of Theorem Ab3, yields:

Theorem Ab7: For all subspaces E of $T^{(2)}$,

$$d_E = o(\log_m(\dim E)), \quad (m = 1, 2, \ldots).$$

(The same result holds in $(T^{(2)})^*$.

We introduce some notation for the next result.

Definition Ab8: If $\{x_i\}_{i=1}^n$ is a sequence in a Banach space X, the unconditional basis constant of $\{x_i\}_{i=1}^n$, denoted "U.B.C. $\{x_i\}$", is the smallest $\lambda > 1$ such that

$$\|\sum_{i=1}^n \varepsilon_i \alpha_i x_i\| \leq \lambda \|\sum_{i=1}^n \alpha_i x_i\|,$$

for all scalar choices $\{\alpha_i\}$ and all choices of signs $\{\varepsilon_i\}$. In this case, we call $\{x_i\}_{i=1}^n$ a λ-unconditional basic sequence.

The proof of Theorem Xe6 is "finite dimensional", and yields the following:

Theorem Ab9: There is a function $f(x,y)$ such that for each $\delta > 0$ and every λ-unconditional basic sequence $\{x_i\}_{i=1}^n$ in $T^{(2)}$ (or its dual), there is a subset F of $\{1, 2, \ldots, n\}$ so that

(i.) $|F| \geq \delta n$, and

(ii.) $\{x_i : i \epsilon F\}$ is $f(\lambda, \delta)$ equivalent to the unit vector basis of $\ell_2^{\dim F}$.

Notes and Remarks:

1/. Since $T^{(2)}$ is of type 2, Maurey's projection theorem (see Aa6) shows that the spaces ℓ_2^n are uniformly complemented in $T^{(2)}$. As we mentioned in the last section, it's not known if the converse of this theorem holds. The most likely place to look for a counter-example is in $(T^{(2)})^*$. Since this space is of cotype 2 and contains no Hilbert spaces, it has no infinite dimensional subspaces of type 2. (In fact, it has no subspaces of quotient spaces of type 2.) It is unknown if $(T^{(2)})^*$ has Maurey's extension property or Maurey's projection property.

2/. We will see later that symmetric convexified Tzirezson's space $S(T^{(2)})$, as defined in chapter X, fails Theorem Ab7. However, the following variant holds:

Proposition Ab10: For $E_n := [t_i]_{i=1}^n$ in $S(T^{(2)})$, we have

$$d_{E_n} = o(\log_m(\dim E_n)), \quad (m = 1, 2, \ldots).$$

Proof: Since $T^{(2)}$ is of type 2, for any $x = \sum_{i=1}^n a_i t_i \in E_n$, by Lemma Xb4 and Remark 6 of Xe, there exists a universal constant $K > 1$ such that $\|x\|_{S(T^{(2)})} \leq K\|Dx\|_{T^{(2)}} \leq K\left(\sum_{k=1}^n a_i^2\right)^{1/2}$, where D is the non-increasing rearrangement operator. Conversely, if $F_n := [t_i]_{i=1}^n$ in $T^{(2)}$, $\|x\|_{S(T^{(2)})} \geq \frac{1}{K}\|Dx\|_{T^{(2)}} \geq \frac{1}{Kd_{F_n}}\left(\sum_{i=1}^n a_i^2\right)^{1/2}$. Hence, $d_{E_n} = o(\log_m(\dim F_n)) = o(\log_m(\dim E_n))$. $\square$

3/. Since $\|\sum_{i=1}^n t_i\|_{S(T^{(2)})}$ is K-equivalent to $\sqrt{n}$, it follows by the "Levy-mean" (see [18]) that there are constants $K > 0$ and $c > 0$ such that for each n, there exists a subspace F_n of $[t_i]_{i=1}^n$ in $S(T^{(2)})$ such that $\dim F_n \geq cn$ and $d_{F_n} \leq K$. i.e, $[t_i]_{i=1}^n$ has a Hilbert subspace which is a "percentage" of its dimension. We will see later that these subspaces cannot be uniformly complemented. We will also later note that there exists a sequence $\{E_n\}_{n=1}^{n}$ of subspaces of $S(T^{(2)})$ with $\dim E_n = n$ such that $\ell_2^{[[\frac{n}{2}]]}$ does not embed uniformly into E_n. (This will merely be the statement that $S(T^{(2)}$ is not of weak cotype 2).

Ac: Weak type and weak cotype

In basic work concerning "weak-type 2" and "weak-cotype 2" properties, G. Pisier [26] introduced the definition of weak Hilbert space and developed the properties of these spaces. This work built upon the earlier paper of V. Milman and G. Pisier [17], where these "weak" notions were introduced.

To see that a weak Hilbert space is truly a "weakening" of the definition of a Hilbert space, we need to formulate an equivalent notion of type 2 and cotype 2.

Definition Ac1:

(i.) If X is a Banach space, and $U : \ell_2^n \to$ is an operator, then the ℓ-norm of U is

$$\ell(U) := \left(\int_{R^n} \|U(x)\|^2 \gamma_n(dx) \right)^{1/2} , \text{ where}$$

γ_n is the canonical Gaussian probability measure on R^n.

(ii.) Dually, if $V : X \to \ell_2^n$, we define

$$\ell^*(V) := \sup\{tr(UV) : U : \ell_2^n \to X, \quad \ell(U) \leq 1\}.$$

(iii.) If X, Y are Banach spaces, and $T : X \to Y$ is an operator, for each n, we define the nth approximation number of T as:

$$a_n(T) := \inf\{\|T - S\| : S : X \to Y, \text{ rank } (S) \leq n\}.$$

(iv.) If the operator $T : X \to Y$ factors through a Hilbert space H, the norm of factorization is given by

$$\gamma_2(T) := \inf\{\|U\| \cdot \|V\| : T = UV, U : X \to H, V : H \to Y\}.$$

(v.) An operator $U : X \to Y$ is 2-absolutely summing (or just "2-summing") if there is a constant M such that for all finite sequences $\{x_i\}_{i=1}^{n}$ in X, we have:

$$\left(\sum_{i=1}^{n} \|Ux_i\|^2 \right)^{\frac{1}{2}} \leq M \cdot \sup\left\{ \left(\sum_{i=1}^{n} |f(x_i)|^2 \right)^{1/2} : f \in B_{X^*} \right\}.$$

We denote by $\Pi_2(U)$ the smallest such M, and by $\Pi_2(X, Y)$ the set of all such operators.

It is well known that:

(a) $\ell(U) \leq \Pi_2(U)$, for all $U : \ell_2^n \to X$, and

132

(b) $\Pi_2(V) \leq \ell^*(V)$, for all $V : X \to \ell_2^n$.

Now we can reformulate the definitions of type 2 and cotype 2.

Theorem Ac2:

1. A Banach space X is of cotype 2 iff there is an $M > 0$ such that for all n and all operators $U : \ell_2^n \to X$, we have: $\Pi_2(U) \leq M \cdot \ell(U)$.

2. A Banach space X is of type 2 iff there is an $M > 0$ such that for all n and all operators $V : X \to \ell_2^n$, we have: $\Pi_2(V^*) \leq M \cdot \ell^*(V)$.

The smallest constants for which Theorem Ac2 holds are called the Gaussian cotype 2 and type 2 constants of U. G. Pisier [24] has shown that Theorem Ac2 is equivalent to:

(1') $C_2(X) < \infty$, and

(2') $T_2(X) < \infty$.

The final notion we need for studying weak Hilbert spaces is K-convexity. This idea arose naturally in some work of B. Maurey and G. Pisier [16] dealing with the duality between type and cotype:

Definition Ac3: Let D be the set $\{-1, +1\}^N$ and $\varepsilon_n : D \to \{-1, +1\}$ be the nth coordinate function. Let μ be the normalized Haar measure on the compact group D, and let R_1 be the orthogonal projection from $L^2(D, \mu)$ onto the closed linear span of $\{\varepsilon_n : n\varepsilon N\}$. A Banach space X is K-convex if the operator $R_1 \otimes I_x$, defined on $L^2(D, \mu) \otimes X$, extends to a bounded operator from $L^2(D, \mu, X)$ into itself. The extension constant is denoted $K(X)$ and called the K-convexity constant of X.

In [15] it was shown:

Theorem Ac4: If a Banach space X is K-convex, then X is of type p iff X^* is of cotype q, for $\frac{1}{p} + \frac{1}{q} = 1$.

G. Pisier [25] showed:

Theorem Ac5: A Banach space X is K-convex iff X does not contain the spaces ℓ_1^n uniformly.

So X is K-convex iff X is of type p, for some $p > 1$. Also, it's a well known result in the theory of type and cotype that X is K-convex iff X^* is K-convex, and that $K(X) = K(X^*)$. (See [24]). Moreover, T. Figiel and N. Tomczak-Jaegerman [7] have shown:

Theorem Ac6: A Banach space X is K-convex iff there is an $M > 0$ such that for all n and all operators $V : X \to \ell_2^n$, we have $\ell(V^*) \leq M\ell^*(V)$. Moreover, $\ell(V^*) \leq K(X)\ell^*(V)$.

Finally, we're in position to introduce weak type and weak cotype.

Definition Ac7:

1. A Banach space X is a weak cotype 2 space if there is a constant $M > 0$ so that for all n and all operators $U : \ell_2^n \to X$, we have:

$$\sup \sqrt{k} \cdot a_k(U) \leq M \cdot \ell(U).$$

2. A Banach space X is a weak type 2 space if there is a constant $M > 0$ so that for all n and all operators $V : X \to \ell_2^n$, we have

$$\sup \sqrt{k} \cdot a_k(V) \leq M \cdot \ell^*(V).$$

The smallest M in (1) is called the weak cotype 2 constant of X, (denoted "$wC_2(X)$") and similarly we define the weak type 2 constant of X, "$wT_2(X)$".

Now let $U : \ell_2^n \to X$ be an operator. If $\{x_i\}_{i=1}^n$ is an orthonormal basis of ℓ_2^n, and if X is of type 2, then Theorem Ac2 implies $\ell(U) \leq C \left(\sum_{i=1}^n \|Ux_i\|^2 \right)^{1/2}$. By duality, this is equivalent to : for every $V : \ell_2^n \to X$,

$$\left(\sum_{i=1}^n \|V^*x_i\|^2 \right)^{1/2} \leq C\ell^*(V).$$

Then by [22],

$$\left(\sum_{i=1}^n a_i(V)^2 \right)^{1/2} = \left(\sum_{i=1}^n a_i(V^*)^2 \right)^{1/2} \leq C\ell^*(V).$$

Since $a_1(V) \geq a_2(V) \geq \ldots, k^{\frac{1}{2}}a_k(V) \leq \sum_{i=1}^k \frac{a_i(V)}{k^{1/2}} \leq \left(\sum_{i=1}^k a_i(V)^2 \right)^{1/2}$, and so $\sup_{k \geq 1} k^{1/2}a_k(V) \leq C\ell^*(V)$, and we now have:

Proposition Ac8: If X is type 2, (respectively, cotype 2), then X is weak type 2 and $wT_2(X) \leq T_2(X)$, (respectively, $wC_2(X) \leq C_2(X)$).

It follows immediately that the wT_2 property is stable with respect to subspaces, quotients, and finite direct sums. In addition, any space which is finitely representable in a wT_2 space is also wT_2. In order to classify wC_2 spaces we need several results from V. Milman and G. Pisier [17].

Lemma Ac9: Let X be a Banach space and $U : \ell_2^k \to X$ be an operator. Then, for any m, there exists a subspace E of ℓ_2^k with $\dim E > k - m$ such that $\|U|_E\| \leq m^{-1/2}d_X\ell(U)$.

Proof: A standard induction argument yields an orthonormal basis $\{x_i\}_{i=1}^k$ for ℓ_2^k such that $\|Ux_i\| \geq a_i(U)$, $(1 \leq i \leq k)$. We now have $\left(\sum_{i=1}^k a_i(U)^2 \right)^{1/2} \leq \left(\sum_{i=1}^k \|Ux_i\|^2 \right)^{1/2} \leq d_X\ell(U)$. Hence $a_m(U) \leq m^{-1/2}d_X\ell(U)$, but this is the conclusion of the Lemma. $\square$

The next step in classifying cotype 2 spaces is also due to V. Milman and G. Pisier [17]:

Proposition Ac10: Let X be a Banach space and assume that there exist constants $0 < \delta_o < 1$ and $c(\delta_o) > 0$ such that for any finite dimensional subspace E of X there exists a subspace F of

E such that $d_F \leq c(\delta_o)$ and $\dim F \geq \delta_o \dim E$. Then there exists a constant $c = c(X) > 0$ such that for each n and every operator $U : \ell_2^n \to X$ there exists a subspace G of ℓ_2^n with $\mathrm{codim}\, G \leq k$ such that $\|U|_G\| \leq ck^{-1/2}\ell(U)$.

Proof: Let $U : \ell_2^n \to X$. Without loss of generality, assume $\ker U = 0$. Let $\alpha = \frac{\delta_o}{2}$ and $E = U(\ell_2^n)$. Then there exists a subspace $F \subset E$ with $\dim F \geq \delta_o \dim E = \delta_o n$ and $d_F \leq c(\delta_o)$. (By adjusting our constants a bit, we may assume that $n = \frac{1}{\alpha^m}$, for some m, and that $\delta_o n$ is an integer.) By Lemma Ac9, there exists a subspace $G_1 \subset \ell_2^n$ with $\dim G_1 = \alpha\, n$ such that $\|U|_{G_1}\| \leq c(\delta_o)(\alpha\, n)^{-1/2}\ell(U)$. Since $\dim G_1^\perp = (1 - \alpha)n$, we can repeat the construction with $G_1^\perp$ in lieu of ℓ_2^n. Then there exists a subspace $G_2 \subset G_1^\perp$ with $\dim G_2 = \alpha\,(1 - \alpha)n$ and

$$\|U|_{G_2}\| \leq c(\delta_o)\ell(U)(\alpha\,(1-\alpha)n)^{-\frac{1}{2}}.$$

After t steps we produce orthogonal subspaces $G_1, \ldots, G_t$ with
$\sum_{i=1}^{t} \dim G_i = (1 - (1 - \alpha)^t)n$, and $\|U|_{G_i}\| \leq c(\delta_o)\ell(U)[\alpha\,(1-\alpha)^{i-1}n]^{-1/2}, (i = 1, 2, \ldots, t)$. Thus,

$$\|U|_{\sum_{i=1}^{t} G_i}\| \leq \left(\sum_{i=1}^{t}\|U|_{G_i}\|^2\right)^{1/2} \leq \alpha^{-1/2}\, c(\delta_o)\ell(U)n^{-1/2}\left(\sum_{i=0}^{t-1}(1-\alpha)^{-i}\right)^{1/2} \leq$$

$\alpha^{-1/2}\, c(\delta_o)\ell(U)n^{-1/2}(1-\alpha)^{-(\frac{t-1}{2})}$, so if $k \leq n$ and $k_t = \mathrm{codim}\sum_{i=1}^{t} G_i = (1-\alpha)^t n$, then by choosing the smallest such t and letting $G = \sum_{i=1}^{t} G_i$, we have $\mathrm{codim}\, G < k$ and $\|U|_G\| \leq c(\delta_o)\,\alpha^{-1}\ell(U)k^{-1/2}$. $\square$

Proposition Ac10 shows that $\sup_{k>1} k^{1/2}a_k(U) \leq c\ell(U)$, for spaces satisfying its hypotheses.
The next result is also from [17].

Theorem Ac11: For a Banach space X, the following are equivalent:

1. There is a $0 < \delta_o < 1$ and a $c(\delta_o) > 1$ such that for every finite dimensional subspace $E \subset X$, there exists a subspace $F \subset E$ with $\dim F \geq \delta_o \dim E$ and $d_F \leq c(\delta_o)$.

2. There is a $0 < \delta_o < 1$ and a $c(\delta_o) > 0$ such that for each n and each operator $U : \ell_2^n \to X$ we have

$$a_{[[\delta_o n]]}(U) \leq c(\delta_o)n^{-1/2}\ell(U)$$

3. There is a constant c such that for all compact operators $U : \ell_2 \to X$ we have:

$$\sup_k k^{1/2}a_k(U) \leq \ell(U).$$

Moreover, these properties imply:

4. There is a constant c such that for any finite sequence $\{x_i\}_{i=1}^{n}$ in X for which $\sup_{i \leq n}|a_i| \leq$

$\|\sum_{i=1}^{n} a_i x_i\|$ for all sequences $\{a_i\}_{i=1}^{n}$ of scalars, then

$$\sqrt{n} \leq c \left(\int_{R^n} \|\sum_{i=1}^{n} a_i x_i\|^2 \gamma_n(d\propto) \right)^{1/2}.$$

Proof: The proof of Proposition Ac10 includes (1) $\Rightarrow$ (2) $\Rightarrow$ (3). To see (3) $\Rightarrow$ (1), let E be a subspace of X with $\dim E = n$. Then there exists an isomorphism $U : \ell_2^n \to E$ such that $\|U\| \leq 1$ and $\Pi_2(U^{-1}) \leq \sqrt{n}$. Hence, for any subspace $G \subset \ell_2^n$ with $\dim G > n - k$, we have

$$\sqrt{n-k} = \Pi_2(Id_G) \leq \|U|_G\|\Pi_2(U^{-1})$$
$$\leq \sqrt{n}\|U|_G\|.$$

Thus, $\|U|_G\| \geq \left(1 - \frac{k}{n}\right)^{1/2}$, and so $a_k(U) \geq (1 - \frac{k}{n})^{1/2}$. If $k = [[\frac{n}{2}]]$, by (3) we obtain:

$$\ell(U) \geq c^{-1} \left[\left[\frac{n}{2}\right]\right]^{1/2} 2^{-\frac{1}{2}}.$$

Then, by a result of T. Figiel, J. Lindenstrauss, and V. Milman [6], there exists a subspace $F \subset E$ with $\dim F \geq \delta_o n$ and $d_F \leq 2$. (Here, $\delta_o = ac^{-2}$, for a universal a). This establishes (1).

To see that (3) $\Rightarrow$ (4), let $\{x_i\}_{i=1}^{n}$ satisfy the property in (4). Let $E = [x_i]_{i=1}^{n}$ and let $U : \ell_2^n \to E$ be defined by $U(\{a_i\}_{i=1}^{n}) = \sum_{i=1}^{n} a_i x_i$. Then $\Pi_2(U^{-1}) \leq \sqrt{n}$ (since $U^{-1} = iV$, where $i : \ell_\infty^n \to \ell_2^n$ is the inclusion map and $V : E \to \ell_2^n$ satisfies $\|V\| \leq 1$, by (4)).

By (3), there exists a subspace $G \subset \ell_2^n$ with $\dim G = [[\frac{n}{2}]]$ and, if P is the orthogonal projection onto G, $\|UP\| \leq c(2n^{-1})^{1/2}\ell(U)$. Thus,

$$[[\tfrac{n}{2}]] = \dim G = tr(U^{-1}UP)$$
$$\leq \Pi_2(U^{-1})\Pi_2(UP)$$
$$\leq n\|UP\| \leq cn^{\frac{1}{2}}2^{\frac{1}{2}}\ell(U).$$

For large n, $\ell(U) > (4c)^{-1}n^{\frac{1}{2}}$, and so we obtain (4). $\square$

Our next objective is to show that (1) of Theorem Ac11 holds for every $0 < \delta < 1$, (This is also a result of [17].) We begin by recalling some facts about ℓ-norms. The first is a result of D. Lewis [11].

Theorem Ac12: For any n-dimensional space E, there exists an isomorphism $U : \ell_2^n \to E$ such that $\ell(U) = \ell^*(U^{-1}) = \sqrt{n}$.

We also require a result of G. Pisier [25]:

Theorem Ac13: There exists an absolute constant $K > 0$ such that for all $U : \ell_2^n \to E$,

$$\ell(U) \leq \ell^*(U^*)K \log(1 + d_E).$$

Finally we require a result of V. Milman [19].

Theorem Ac14: There is a function $\Psi : (0,1) \to R$ satisfying: for any Banach space X and any operator $V : X \to \ell_2^n$ and any $0 < \epsilon < 1$, there exists a subspace $E \subset X$ with $\text{codim} E < \epsilon n$ for which $\|V|_E\| \leq \Psi(\epsilon) n^{\frac{1}{2}} \ell(V^*)$.

We can now prove the result of V. Milman and G. Pisier [17].

Theorem Ac15: Let X be a Banach space. If there exist constants $0 < \delta_o < 1$ and $c(\delta_o) > 0$ such that for every finite dimensional subspace $E \subset X$ there exists a subspace $F \subset E$ for which $\dim F \geq \delta_o \dim E$ and $d_F \leq c(\delta_o)$, then for every $0 < \delta < 1$ there exists a constant $c(\delta)$ for which this holds.

Proof: Let $o < \delta < 1$ and $\epsilon = 1 - \delta$. Let E be a subspace of X with $\dim E = n$. By Theorem Ac12, there exists an isomorphism $U : \ell_2^n \to E$ with $\ell(U) = \ell^*(U^{-1}) = \sqrt{n}$. By Proposition Ac10, there exists a subspace $H \subset \ell_2^n$ with $\text{codim} H < \frac{\epsilon n}{2}$, and

$$(1) \qquad \|U|_H\| \leq c\left(\frac{2}{\epsilon}\right)^{1/2} n^{-1/2} \ell(U).$$

Let $|\cdot|$ denote the Euclidian norm on ℓ_2^n. Fix $p > 0$ and define a new norm on E by:

$$\|x\|_p = \|x\| + p|U^{-1}x|, \qquad (x \epsilon E).$$

Let $E_1 = U(H) \subset E$. Then, for each $x \epsilon E_1$, by (1),

$$(2) \qquad p|U^{-1}| \leq \|x\|_p \leq \left(c(\frac{2}{\epsilon})^{1/2} + p\right)|U^{-1}x|.$$

Let E_1^p denote E_1 with the norm $\|\cdot\|_p$. By (2),

$$d_{E_1^p} \leq 1 + c(\frac{2}{\epsilon})^{1/2}p^{-1}.$$

Let $j : E_1^p \to E_1$ be the identity. Then

$$\|j\| \leq 1, \text{ and } \ell^*(U^{-1}j) \leq \ell^*(U^{-1}) = \sqrt{n}.$$

By Theorem Ac13,

$$\ell((U^{-1}j)^*) \leq B_p\sqrt{n}, \qquad \text{where}$$
$$B_p = K\log(2 + c(\frac{2}{\epsilon})^{1/2}p^{-1}).$$

By Theorem Ac14, there exists a subspace $F \subset E_1^p$ with $\text{codim} F \leq \frac{\epsilon n}{2}$ such that $\|U^{-1}j|_E\| \leq \Psi(\frac{\epsilon}{2})B_p$. Thus, for all $x \in F$,

$$\left(\Psi\left(\frac{\epsilon}{2}\right)B_p\right)^{-1}|U_x^{-1}| \leq \|x\|_p.$$

Hence, $\left(\Psi(\frac{\epsilon}{2})B_p\right)^{-1}\left\{1 - pB_p(\frac{\epsilon}{2})\right\}|U^{-1}x| \leq \|x\|$. Since $pB_p \to 0$, as $p \to 0$, we may choose $p = F(\delta)$ such that $pB_p\Psi(\frac{\epsilon}{2}) = \frac{1}{2}$. Then,

$$(3) \qquad p|U_x^{-1}| \leq \|x\|, \qquad (x \in F).$$

137

But, since $F \subset E_1^p$, by (1) again, we obtain

(4)
$$\|x\| \leq c(\frac{2}{\epsilon})^{\frac{1}{2}}|U^{-1}x|, \qquad (x \in F).$$

Finally, upon considering F as a subspace of E, and letting $\hat{F}$ denote the cooresponding normed space, (3) and (4) imply

$$d_{\hat{F}} \leq p^{-1}c\left(\frac{2}{\epsilon}\right)^{1/2}.$$

Also, $\dim \hat{F} = \dim F \geq \dim E_1 - \frac{\epsilon n}{2}$, so $\dim \hat{F} \geq n - \epsilon n = \delta n$. $\quad\square$

In proving the above, V. Milman and G. Pisier produced good estimates for the constants involved. In particular, they showed $c(\delta) \leq c'(1-\delta)^{-1}\log(c'(1-\delta)^{-1})$, where c' is a constant depending only upon δ_o and $c(\delta_o)$.

Our next result shows that wT_2 is "close" to type 2.

Theorem Ac16: A Banach space X is wT_2 iff X^* is K-convex and wC_2. Moreover, $wC_2(X^*) \leq wT_2(X) \leq K(X)\,wC_2(X^*)$.

Proof: By Theorem Ac6, for all operators $V : X \to \ell_2^n$,

$$\ell(V^*) \leq K(X)\ell^*(V).$$

So if X^* is K-convex and wC_2, then

$$\sup_k k^{1/2}a_k(V^*) \leq \ wC_2(X^*)\ell(V^*) \leq \ wC_2(X^*)K(X)\ell^*(V).$$

Hence, X is $wT_2(X) \leq \ wC_2(X^*)K()$.

If X is wT_2, then X is of type p, for some $p > 1$. (We need only mimic the proof that X type p implies X is K-convex to see this.) So X is K-convex. For any operator $V : X \to \ell_2^n, \ell^*(V) \leq \ell(V^*)$, so X^* is wC_2 and $wC_2(X^*) \leq \ wT_2(X)$. $\quad\square$

Our next characterization of wT_2 is useful in applications. In its proof we will need the following "lifting" lemma of G. Pisier [23].

Lemma Ac17: Let E be a closed subspace of a K-convex space X. Let $Q : X \to X/E$ be the quotient map. For any operator $V : \ell_2^n \to X/E$, there exists a lifting $\hat{V} : \ell_2^n \to X$ such that $Q\hat{V} = Q$ and $\ell(\hat{V}) \leq 2K(X)\ell(V)$.

The next theorem is due to V. Milman and G. Pisier [17].

Theorem Ac18: For a Banach space X, the following are equivalent:

1. X is wT_2.
2. For all $0 < \delta < 1$, there exists $c(\delta) > o$ such that for any subspace $E \subset X$, any n, and any operator $U : E \to \ell_2^n$, there exists an orthogonal projection $P : \ell_2^n \to \ell_2^n$ with $\text{rank}P \geq \delta n$ and an extension $\hat{U} : X \to \ell_2^n$ of PU (i.e., $\hat{U}|_E = PU$) such that $\|\hat{U}\| \leq c(\delta)\|U\|$.

3. Same as (2), except that there exists one $0 < \delta < 1$ and a $c(\delta) > 1$ satisfying (2).

Proof: To see that $(1) \Rightarrow (2)$, Let $U : E \to \ell_2^n$. Then $U^* : \ell_2^n \to X^*/E^\perp$. Let $Q : X^* \to X^*/E^\perp$ be the quotient map. Lemma Ac16 implies that there exists an operator $\hat{V} : \ell_2^n \to X^*$ such that $Q\hat{V} = U^*$ and $\ell(\hat{V}) \le 2K(X)\ell(\hat{V})$. Since $\ell(U^*) \le n^{\frac{1}{2}}\|U\|$, for any U, we obtain

$$\ell(\hat{V}) \le 2n^{1/2}K(X)\|U\|.$$

Let $\hat{U} = \left(\hat{V}|_X\right)^*$. Then $\hat{U} : X \to \ell_2^n$ and $\hat{U}|_E = U$, (and $\hat{U}^* = \hat{V}$), so

$$\ell(\hat{U}) \le 2n^{\frac{1}{2}}\|U\|K(X).$$

Since X^* is wC_2 (and $a_k(\hat{U}^*) = a_k(\hat{U})$), we have $\sup_k k^{\frac{1}{2}}a_k(\hat{U}) \le wC_2(X^*)\ell(\hat{U}^*)$. Hence, for all $1 < k < n$, there exists a projection P on ℓ_2^n with $\mathrm{rank}P > n - k$ such that

$$\|P\hat{U}\| \le wC_2(X^*)\ell(\hat{U}^*)k^{-\frac{1}{2}}.$$

Adjusting k gives the result. That $(2) \Rightarrow (3)$ is obvious. To see that $(3) \Rightarrow (1)$, clearly one notes that $(3) \Rightarrow X$ is K-convex (e.g., the proof of Notes and Remarks 2, of section Aa.) Let $U : \ell_2^n \to X^*$. By Theorem Ac14, there exists a subspace $E \subset X$ with $\mathrm{codim}E < \frac{\delta n}{2}$ and $\|U^*|_E\| \le c\ell(U)n^{-\frac{1}{2}}$. By (3), there exists a projection $P : \ell_2^n \to \ell_2^n$ with $\mathrm{rank}P \ge \delta n$ and an operator $V : X^* \to \ell_2^n$ such that $V|_E = PU^*|_E$ and $\|V\| \le c\|U^*|_E\|$. Since $(PU^* - V)|_E = 0$, we obtain $\mathrm{rank}(PU^* - V) < \frac{\delta n}{2}$. Now, if $T = U^* - V$, then

$$\mathrm{rank}T < \mathrm{rank}(PU^* - V) + \mathrm{rank}(I - P) < (1 - \frac{\delta}{2})n.$$

Moreover, $\|U - T^*\| = \|V\| \le c_o c\ell(U)n^{\frac{1}{2}}$. Hence,

$$a_k(U) \le c_o c\ell(U)n^{-\frac{1}{2}}, \quad \text{for } k = [[(1 - \frac{\delta}{2})n]].$$

Theorem Ac11 now implies X^* is wC_2 and (since X^* is K-convex) so X is wT_2. $\quad \square$

It follows that if X is wT_2, then $\sup\{p|X$ is type $p\} = 2$. Hence, if X is wC_2, then $\inf\{q|X$ is cotype $q\} = 2$.

We also have the following from [17]:

Theorem Ac19: Let $1 \le p \le 2$. A Banach space X is of cotype 2 iff $L_p(X)$ is of weak cotype 2.

Proof: We need only demonstrate the "if" part.) If $L_p(X)$ has wC_2, then for all $\{x_i\}_{i=1}^n \subset X$ with $\|x_i\| \ge 1, (1 \le i \le n)$ and all scalar choices $\{a_i\}_{i=1}^n$,

$$\sup|a_i| \le \|\sum_{i=1}^n a_i r_i x_i\|_{L_p(X)}.$$

Hence, by (4) of Theorem Ac11,

$$\sqrt{n} \le c\left(\int\|\sum_{i=1}^n a_i r_i x_i\|_{L_p(X)}^2 d\gamma_n(a)\right)^{1/2}$$
$$\le c\left(\int\|\sum_{i=1}^n a_i r_i x_i\|_{L_2(X)}^2 d\gamma_n(a)\right)^{/2}$$

By symmetry and homogeneity, this implies that

$$\sqrt{n}\inf_{1\le i\le n}\|x_i\| \le c\left(\int\|\sum_{i=1}^{n}a_ix_i\|^2 d\gamma_n(a)\right)^{1/2}.$$

But, by a result of R.C. James [8], this implies X is cotype 2. $\square$

With the notation of definition Ac3, we denote by $\mathrm{Rad}_n X$ the subspace of $L_2(X)$ spanned by $\left\{\sum_{i=1}^{n}\epsilon_ix_i|x_i \in X\right\}$. We denote by $\mathrm{Rad}\,(X)$ the closure on $L_2(X)$ of $\bigcup_{u=1}^{\infty}\mathrm{Rad}_n(X)$ It now follows that if $\mathrm{Rad}X$ is wC_2, then X is cotype 2. Dually, if $\mathrm{Rad}X$ is wT_2, then X is type 2.

Notes and Remarks:

1/. The notions wT_2 and wC_2 have natural extensions to $wT_p,(1 < p < 2)$, and $wC_p,(p > 2)$. In particular, we have:

Definition Ac20: A Banach space X is weak cotype q, $(2 \le 1 < \infty)$, if there exists a constant $c > 0$ such that

$$\sup_k k^{\frac{1}{q}}a_k(U) \le c\ell(U), \quad (U : \ell_2^n \to X).$$

(Similarly, we can define weak type $p,(1 \le p \le 2)$).

Mascioni and Matter [20] have recently proved the following interesting result:

Theorem Ac21: A Banach space X is

1. equal norm type $p,(1 < p < 2)$, iff X is wT_p.
2. equal norm cotype $q,(2 < q < \infty)$, iff X is wC_q.

Recall that in section XE we observed that the Tsirelson space $T^{(p)},(2 < p < \infty)$, is equal norm cotype p but not cotype p. Now we know that it is actually wC_p.

2/. Another area of research which comes into play in weak Hilbert spaces is the notion of volume ratios. Ellipsoids of maximum volume were introduced by F. John [29] in 1948, and became an object of renewed study by S.J. Szarek [27, 28] thirty years later, as a tool in studying decompositions of Banach spaces into ℓ_2^n, s.

The notion can be approached as follows:

Definition Ac22: If $(E, \|\cdot\|)$ is a finite-dimensional Banach space, $n := \dim E$, then every isomorphism $U : \ell_2^n \to E$ induces an inner product $[\cdot,\cdot]$ on E via:

$$[x,y] := \left(U_x^{-1}, U_y^{-1}\right), \qquad (x,y \in E).$$

The Euclidean norm $|\cdot|_2$ on E is defined by:

$$|x|_2 := [x,x]^{1/2}, \qquad (x \in E).$$

The ellipsoid $\varepsilon := \{x \in E : |x|_2 \le 1\}$ is just $U(B_2)$, where B_2 is the unit ball of ℓ_2^n. If we denote by "$\mathrm{vol}_E U(B_2)$" the volume of $U(B_2)$ in the sense of n-dimensional Lebesque measure on the

140

Borel subsets of E, then F. John showed that there is such an ellipsoid of maximal volume. That is, there is a $U_o : \ell_2^n \to E$ so that

$$\varepsilon = U_o(B_2) \subset B_E, \text{ and}$$

$$\text{vol}\varepsilon = \sup\{\text{vol}_E U(B_2) : U : \ell_2^n \to E, U(B_2) \subset B_E\}.$$

Similarly, there is an ellipsoid of minimal volume containing B_E, and both these ellipsoids are unique [see 28]. The volume ratio of E, denoted "$\text{vr}E$", is given by:

$$\text{vr}E := \left(\frac{\text{vol}B_E}{\text{vol}\varepsilon}\right)^{1/n},$$

where $n := \dim E$, and ε is the elliipsoid of maximal volume contained in B_E.

V. Milman and G. Pisier [17] have shown:

Theorem Ac23: A Banach space X is wC_2 iff $\sup\{\text{vr}E | E \subset X, \dim E < \infty\} < \infty$.

Ad: Weak Hilbert Spaces

Finally, we can state:

Definition Ad1: A Banach space X is a weak Hilbert space if it is both wT_2 and wC_2.

The results of the last section immediately entail:

Theorem Ad2: For a Banach space X, the following are equivalent:

1. X is a weak Hilbert space.
2. For every $0 < \delta < 1$, there exists a $c(\delta) > 1$ such that every finite dimensional subspace $E \subset X$ contains a subspace F with $\dim F \geq \delta \dim E, d_F \leq c(\delta)$, and there exists a projection $P : X \to F$ with$\|P\| \leq c(\delta)$.
3. There exists $0 < \delta_o < 1$ and $c(\delta_o) > 1$ such that (2) holds.

G. Pisier (using the groundwork of section Ac) goes on to show [26]:

Theorem Ad3: For a Banach space X, the following are equivalent:

1. X is a weak Hilbert space.
2. There exist constants c and $0 < \alpha < 1$ such that for all n and all operators $U : \ell_2^n \to X$, we have

$$a_{[[\alpha n]]}(U) \leq c[[\alpha \, n]]^{1/2}\Pi_2(U^*).$$

3. There exists a constant c such that for all n and all $U : \ell_2^n \to X$ we have

$$\sup_{k \leq n} k^{\frac{1}{2}}a_k(U) \leq c\Pi_2(U^*).$$

4. There exists a constant c such that any operator $U : H \to X$ from a Hilbert space into X
with a 2-absolutely summing adjoint satisfies

$$\sup_k k^{1/2} a_k(U) \leq c\Pi_2(U^*).$$

Proposition Ab2 and Theorem Ad2 yield immediately:

Theorem Ad4: The spaces $T^{(2)}$ and $(T^{(2)})^*$ are weak Hilbert spaces.

In particular, $T^{(2)}$ is type 2 and wC_2, while $(T^{(2)})^*$ is cotype 2 and wT_2. Since ℓ_2 embeds into
neither $T^{(2)}$ nor $(T^{(2)})^*$, it follows that $T^{(2)}$ is wC_2 but not cotype 2, while $(T^{(2)})^*$ is wT_2 but
not type 2.

Weak Hilbert spaces are stable under subspaces, duals, quotients, and ultraproducts (any
space finitely representable in a weak Hilbert space is also a weak Hilbert space). G. Pisier [26]
has shown that they are also stable under interpolation. It also follows that for weak Hilbert
spaces $X, \sup\{p|X$ is type $p\} = \inf\{q|X$ is cotype $q\} = 2$.

Also, by Theorem Ac19, we obtain:

Theorem Ad5: If $\ell_2(X)$ or $\mathrm{Rad}X$ (or $\mathrm{Rad}_n X$, uniformly over n) are weak Hilbert spaces,
then X is a Hilbert space.

Finite dimensional subspaces of a weak Hilbert space are "close" to finite dimensional Hilbert
spaces, as the next result [26] of G. Pisier shows:

Theorem Ad6: If X is a weak Hilbert space, there exists a constant $K = K(X)$ such that,
for all n-dimensional subspaces $E \subset X$, we have $d_E \leq K \log n$.

Proof: Let $\dim E = n$ and $E \subset X$. Let $i : E \to X$ be the inclusion map. Assume $T : E \to E$
and that $\gamma_2^*(iT) < 1$. Then there exists an operator $\hat{T} : X \to X$ which extends T such that
$\gamma_2^*(\hat{T}) < 1$. Let $\hat{T} = BA$ be a factorization of $\hat{T}$ through a Hilbert space. By Theorem Ad3,
There exists a constant c for which

$$a_k(A^*) = a_k(A) \leq ck^{-\frac{1}{2}}\Pi_2(A), \text{ and}$$
$$a_k(B) \leq ck^{-\frac{1}{2}}\Pi_2(B^*), \text{ both for } k = 1, 2, \ldots.$$

Thus
$$a_{2k}(\hat{T}) \leq a(A)a_k(B) \leq c^2 k^{-1}\Pi_2(A)\Pi_2(B^*), \text{ and so}$$
$$a_{2k}(\hat{T}) \leq c^2 k^{-1}\gamma_2^*(T).$$

Thus
$$\sup na_n(\hat{T}) \leq 4c^2\gamma_2^*(\hat{T}). \quad \text{By G. Pisier [26],}$$

we know
$$\sup n|\lambda_n(\hat{T})| \leq K \sup na_n(\hat{T}),$$

where $\left\{\lambda_n(\hat{T})\right\}$ is the sequence of eigenvalues of $\hat{T}$ in decreasing order (reported according to multiplicities). It folows that $\lambda_k(\hat{T}) \leq ck^{-1}$, of all k, and so

$$
\begin{aligned}
|trT| &= |\sum_{k=1}^{n} \lambda_k(T)| \leq \sum_{k=1}^{n} |\lambda_k(T)| \\
&\leq \sum_{k=1}^{n} |\lambda_k(\hat{T})| \leq c \sum_{k=1}^{n} \frac{1}{k} \leq c \log n
\end{aligned}
$$

Hence, $|trT| \leq K \log n \gamma_2^*(i\,T)$. By the Hahn-Banach Theorem, there exists an operator P : $X \to E$ for which $P|_E = I|_E$ and $\gamma_2(P) \leq K \log n$. Clearly P is a projection onto E, and $d_E \leq \gamma_2(P) \leq K \log n$. $\quad\square$

In $T^{(2)}$ and its dual, we have a much stronger estimate (see Theorem Ab7). It is not known if a stronger estimate exists in general for weak Hilbert spaces. Since $S(T^{(2)})$ satisfies this estimate (Proposition Ab10) and is not a weak Hilbert space, this estimate does not classify weak Hilbert spaces.

In V. Milman and G. Pisier [17], the following characterization of wC_2 spaces is given (see Notes and Remarks of section Ac for notation):

Theorem Ad7: In a Banach space X, the following are equivalent:

1. X is wC_2.

2. $\sup \{vr(E)|E \subset X, \dim E < \infty\} < \infty$.

A. Pajor [21] improved this result and showed:

Theorem Ad8: A Banach space X is wT_2 iff there exists a $c > 0$ such that for all finite-dimensional quotient spaces Q of X^*, we have $vr(Q) \leq c$.

In [26], G. Pisier uses this to classify weak Hilbert spaces.

Theorem Ad9: A Banach space X is a weak Hilbert space iff there exists a constant $c > 1$ such that for any n and any n-dimensional subspace $E \subset$ there exist ellipsoids D_1 and D_2 in E such that $D_1 \subset B_E \subset D_2$ and $\left(\frac{volD_2}{volD_1}\right)^{1/n} \leq c$.

Using this, G. Pisier has given the following characterization of weak Hilbert spaces.

Theorem Ad10: A Banach space X is a weak Hilbert space iff there exists a constant c such that, for all even integers n, every n-dimensional subspace $E \subset X$ admits a ("Kashin") decomposition $E = E_1 + E_2$ with $\dim E_1 = \dim E_2 = \frac{n}{2}$ and so that $d_{E_1} \leq c$, $d_{E_2} \leq c$, and there exist projections $Q_i : X \to E_i$, with $\|Q_i\| \leq c$, $(i = 1, 2)$.

Proof: (The "if" part is clear). By Theorems Ad9 and Ad7 there exist constants c_1, c_2 so that for any n-dimensional subspace $E \subset X$, we can find two Euclidean norms $|\cdot|_1$ and $|\cdot|_2$ on E with associated unit balls D_1 and D_2 such that $D_1 \subset B_E \subset D_2$ and $\left(\frac{volB_E}{volD_1}\right)^{\frac{1}{n}} \leq c_1$, and

$\left(\frac{vol D_2}{vol B_E}\right)^{\frac{1}{n}} \leq c_2$. Moreover, there exists a projection $P : X \to E$ such that

$$|Px|_2 \leq \|x\|, \quad (x \in X).$$

Then $\left(\frac{vol D_2}{vol D_1}\right)^{\frac{1}{n}} \leq c_1 c_2$, and so there exists a decomposition $E = E_1 + E_2$ with $\dim E_1 = \dim E_2 = \frac{n}{2}$ and such that $\frac{1}{K}|x|_1 \leq |x|_2 \leq |x|_1$, for all $x \in E_1 \bigcup E_2$ (where $K = 4\pi c_1 c_2$). Thus,

$$\frac{1}{K}|x|_1 \leq |x| \leq |x|_1, \text{ for all } x \in E_1 \bigcup E_2.$$

Now let P_1, P_2 be the orthogonal projections relative to these Hilbert space norms $|\cdot|_1$ and $|\cdot|_2$ onto E_1 and E_2, respectively. Then, for all $x \in X$, $\|P_1 P x\| \leq |P_1 P x|_1 \leq K|P_1 P x|_2 \leq K|P x|_2 \leq K\|x\|$. Hence, $Q_1 := P_1 P$ is a projection of X onto E_1, with $\|Q_1\| \leq K$. $Q_2 := P_2 P$ is the other projection. $\square$

Theorem Ad10 almost appears to imply that X is a Hilbert space, since each $2n$-dimensional subspace E "splits" into two complemented subspaces $E = E_1 + E_2$. However, if P_1 and P_2 are the respective projections, it does not happen that range $P_1 \subset \ker P_2$, or vice versa, so this decomposition is hardly a direct sum.

We end this section by mentioning one more result of G. Pisier [26].

Theorem Ad11: Every weak Hilbert space has the approximation property (in fact, the uniform approximation property).

The proof is elegant. and well worth the effort required to read it. We have deleted it because it requires still much more theory than is presented here.

Notes and Remarks: 1/. We still have few non-trivial examples of weak Hilbert spaces. Besides $T^{(2)}$, its dual, their direct sum, and interpolations of these, there are almost no examples. We can consider $T^{(2)}(\theta)$ as defined in XA to produce an uncountable family of weak Hilbert spaces which are pairwise totally incomparable, but at the same time they are all "similar".

2/. It is not difficult to show that if $\{t_n\}_{n=1}^{\infty}$ is the canonical unit vector basis for $T^{(2)}$, and $E_n := [t_i]_{i=n}^{\infty}$, then $X := \left(\sum_n \oplus E_n\right)_{\ell_2}$ is a weak Hilbert space.

Ae: Constructibility Properties of weak Hilbert spaces.

Many of the results in this section are due to W.B. Johnson, and appeared for the first time in [18]. We believe that these are among the most important results about weak Hilbert spaces, since they readily lend themselves to constructions inside the space. We begin with several (a priori) weakenings of the notion of Hilbert space:

Definition Ae1:

1. A Banach space X has property H_2 if for every $0 < \delta < 1$ and for every normalized λ-unconditional basic sequence $\{x_i\}_{i=1}^{n}$ in X, there is a subset F of $\{1, 2, \dots, n\}$ such that

$|F| \geq \delta n$ and $\{x_i : i\varepsilon F\}$ is $C(\delta, \lambda)$-equivalent to the unit vector basis of $\ell_2^{\dim F}$ (where $C(\delta, \lambda)$ is a constant depending only on δ and λ).

2. A Banach space X has property H if for each $\lambda > 1$ there is a constant $f(\lambda)$ such that for any n and any normalized λ-unconditional basic sequence $\{x_i\}_{i=1}^n$ in X, we have:

$$(f(\lambda))^{-1} n^{1/2} \leq \|\sum_{i=1}^n x_i\| \leq f(\lambda) \cdot n^{1/2}.$$

3. A Banach space X is said to be asymptotically Hilbertian ("as. Hilbertian") if there is a constant β such that for all n there is a subspace Y_n of finite codimension in X such that each n-dimensional subspace E of Y_n satisfies:

$$d_E \leq \beta.$$

We will next prove that weak Hilbert spaces have property H_2. To simplify this, we begin with a lemma.

Lemma Ae2: For a Banach space X, the following are equivalent:

1. X has property H_2.

2. There exists one $\delta, 0 < \delta < 1$, and a function $C(\delta, \lambda)$ satisfying the conclusion of property H_2.

Proof: ($(1) \rightarrow (2)$ is obvious.) Assume there exists $0 < \delta < 1$ such that, for each λ-unconditional sequence $\{x_i\}_{i=1}^n$ in X, there exists a subset F of $\{1, 2, \ldots, n\}$ such that $|F| \geq \delta n$ and $\{x_i : i \in F\}$ is $C(\delta, \lambda)$-equivalent to the unit vector basis of $\ell_2^{|F|}$. Pick any λ-unconditional basic sequence $\{x_i\}_{i=1}^n$ in X and let $0 < \delta_0 < 1$. By our assumption, there exists an $F_1 \subset E_1 = \{1, 2, \ldots, n\}$ with $|F_1| \geq \delta n$ and $\{x_i : i \in F_1\}$ is $C(\delta, \lambda)$-equivalent to the unit vector basis of $\ell_2^{|F_1|}$. Applying this same assumption to $E_2 = E_1 \backslash F_1$, we assert the existence of an $F_2 \subset E_1 \backslash F_1$ with $|F_2| \geq \delta |E_1 \backslash F_1|$ and $\{x_i : i \in F_2\}$ is $C(\delta, \lambda)$-equivalent to the u.v.b. of $\ell_2^{|F_2|}$. Inductively, we obtain disjoint subsets $F_1, F_2, \ldots, F_m$ of $\{1, 2, \ldots, m\}$ such that $|F_i| \geq \delta[n - \bigcup_{j=1}^{i-1} |F_j|]$, $(i = 2, 3, \ldots, n)$, and $\{x_i : i \in F_j\}$ is $C(\delta, \lambda)$-equivalent to the u.v.b. of $\ell_2^{|F_j|}, (j = 1, 2, \ldots, m)$. It follows that $|\bigcup_{j=1}^m F_j| \geq (1 - \delta) \sum_{j=0}^{m-1} \delta^j$, and so, after a function of δ_0 steps, we find an m with $|\bigcup_{j=1}^m F_j| \geq \delta_0$. Then $\{x_i : i \in \bigcup_{j=1}^m F_j\}$ is $mC(\delta, \lambda)$-equivalent to the u.v.b. of ℓ_2^c for $c = \left| \bigcup_{j=1}^m F_j \right|$. $\square$

Now we can prove:

Theorem Ae3: Every weak Hilbert space has property H_2.

Proof: Let X be a weak Hilbert space, and $\{x_i\}_{i=1}^n$ any λ-unconditional basic sequence in X. We may assume n is even. By Theorem Ad2, there exists a subspace E of $[x_i]_{i=1}^n$ with

$d_E \leq C, |E| = \frac{n}{2}$, and a projection $P : X \to E$ with $\|P\| \leq C$, and C depending only on X (and not on n). Restricting P to $[x_i]_{i=1}^n$ and letting $\{x_i^*\}_{i=1}^n$ be the associated biorthogonal functionals to $\{x_i\}_{i=1}^n$, we have that P induces a decomposition of each x_i into $x_i = y_i + z_i$ (and each x_i^* into $x_i^* = y_i^* + z_i^*$), with $y_i \in E$ and $z_i \in (I|_{[x_i]} - P)E$. If $F = \{1 \leq i \leq n | y_i^*(y_i) \geq \frac{1}{4}\}$, then $\frac{n}{2} = \text{trace} P|_E = \sum_{i=1}^n x_i^*(x_i) \leq \frac{1}{4}|F^c| + \|P\| \, |F|$. But $|F^c| = n$ implies $|F| \geq \frac{1}{4\|P\|} \cdot n$. Applying Lemma VIIa2, we find that $\{x_i\}_{i \in F}$ is $M = M(\lambda, \|P\|) = M(\lambda, C)$-isomorphic to a subspace of $(\sum \oplus E)_{\ell_2}$. But $d_E \leq C$ implies $\{x_i\}_{i \in F}$ is equivalent (up to a function of C and λ) to the u.v.b. of $\ell_2^{|F|}$. Lemma Ae2 now finishes the argument. $\quad\square$

We can now obtain:

Proposition Ae4: Every Banach space with property H_2 also has property H.

Proof: Let $\{x_i\}_{i=1}^n$ be a λ-unconditional basic sequence in X. Fix $\delta = \frac{1}{2}$, and choose $F_1 \subset E_1 := \{1, 2, \ldots, n\}$ such that $|F_1| \geq \frac{n}{2}$ and $\{x_i : i \in F_1\}$ is $f(\lambda) = C(\frac{1}{2}, \lambda)$-equivalent, to the u.v.b. of $\ell_2^{|F_1|}$. Let $E_2 = E_1 \backslash F_1$ and choose $F_2 \subset E_2$ with $|F_2| \geq \frac{|E_2|}{2}$ and $\{x_i : i \in F_2\}$ is $f(\lambda)$-equivalent to the u.v.b. of $\ell_2^{|F_2|}$. Continuing, we obtain disjoint sets $\{F_i\}_{i=1}^m$ with $\bigcup_{i=1}^m F_i = E_1$ and with $\{x_i : i \in F_j\} f(\lambda)$-equivalent to the u.v.b. of $\ell_2^{|F_j|}, (j = 1, \ldots, m)$. With a slight adjustment of n and F_i we many assume $|F_i| = \frac{n}{2^i}, (i = 1, 2, \ldots, m-1)$, and $|F_m| = |F_{m-1}|$. Computing,

$$
\begin{aligned}
\sqrt{n} \ &\leq 2f(\lambda)\|\sum_{i \in F_i} x_i\| \\
&\leq 2\lambda f(\lambda)\|\sum_{i=1}^n x_i\| \\
&\leq 2\lambda f(\lambda)\sum_{j=1}^m \|\sum_{i \in F_j} x_i\| \\
&< 2\lambda (f(\lambda))^2 \sum_{j=1}^m \sqrt{|F_j|} \\
&\leq 2\lambda (f(\lambda))^2 \left(\sum_{j=1}^{m-1} \sqrt{\frac{n}{2^j}} + \sqrt{\frac{n}{2^{m-1}}} \right) \\
&\leq 2\lambda (f(\lambda))^2 \sqrt{n}. \quad \square
\end{aligned}
$$

It is not known whether the converse of Theorem Ae3 (or of Proposition Ae4) holds.

We next show a Banach space with property H must be as. Hilbertian. This is an important result [26] of W.B. Johnson which requires some background.

Definition Ae5:

1. If X is a Banach space, a sequence $\{X_n\}_{n=1}^\infty$ of closed subspaces of X is a Schauder decomposition of X if every $x \in X$ has a unique representation $x = \sum_n x_n$, where $x_n \in X_n$, for every n. In this case, we write $X = \sum \oplus X_n$. If $\dim X_n < \infty$, for all n, we call $\{X_n\}_{n=1}^\infty$ a finite dimensional decomposition ("F.D.D.") for X.

2. A decomposition $\{X_n\}_{n=1}^\infty$ is an unconditional Schauder decomposition ("U.F.D.D.", if each $\dim X_n < \infty$), if for every $x \in X, x = \sum_n x_n$ converges

146

unconditionally.

Every Schauder decomposition $\{X_n\}_{n=1}^{\infty}$ of a Banach space X determines a natural sequence of projections $\{P_n\}_{n=1}^{\infty}$ on X defined via $P_n \sum_i x_i = \sum_{j=1}^{n} x_j$. Then $K := \sup \|P_n\| < \infty$, and K is called the decomposition constant of $\{X_n\}_{n=1}^{\infty}$. If $X = \sum \oplus E_n$ is a F.D.D., to reflect the decomposition constant K, we call $\{E_n\}_{n=1}^{\infty}$ a "K-F.D.D.". If $\{E_n\}_{n=1}^{\infty}$ is an U.F.D.D., then, for every sequence $\theta = \{\theta_n\}_{n=1}^{\infty}$ of signs, the operator M_θ defined by $M_\theta \sum_n x_n = \sum_n \theta_n x_n$ is a bounded linear operator and $K := \sup_\theta \|M_\theta\| < \infty$ and is called the unconditional constant of the decomposition, and we call $\{E_n\}_{n=1}^{\infty}$ a K-U.F.D.D. for X.

Next, we recall the definitions of the Lorentz spaces $\ell_{2\infty}$ and ℓ_{21}. If $x = \{a_n\}_{n=1}^{\infty}$ is a sequence of reals, we denote by by $\{a_n^*\}_{n=1}^{\infty}$ the non-increasing rearrangement of the non-zero elements of $\{|a_n| : n = 1, 3, \ldots\}$. We now define

$$\|x\|_{2\infty} = \sup_{n \geq 1} n^{\frac{1}{2}} a_n^*, \text{ and}$$

$$\|x\|_{21} = \sum_{n=1}^{\infty} \frac{a_n^*}{n^{1/2}}.$$

The Lorentz space $\ell_{2\infty}$ (respectively, ℓ_{21}) is the space of sequences $x = \{a_n\}_{n=1}^{\infty}$ for which $\|x\|_{2\infty} < \infty$ (respectively, $\|x\|_{21} < \infty$). Now we can prove:

Proposition Ae6:

1. If X satisfies property H, then for all normalized λ-unconditional basic sequences $\{x_i\}_{i=1}^{n}$ in X and for all sequences $\{a_i\}_{i=1}^{n}$ of scalars,

$$\frac{1}{\lambda^2 f(\lambda)} \|\{a_i\}_{i=1}^{n}\|_{2\infty} \leq \|\sum_{i=1}^{n} a_i x_i\| \leq \lambda^2 f(\lambda) \|\{a_i\}_{i=1}^{n}\|_{21}.$$

2. If $X = \sum_n \oplus E_n$ is a K-U.F.D.D. for X and X has property H, then for every $x = \sum_n x_n \in X$,

$$\frac{1}{K^2 f(K)} \cdot \|\{\|x_n\|\}_{n=1}^{\infty}\|_{2\infty} \leq \|x\| \leq K^2 f(K) \|\{x_n\}\|_{n=1}^{\infty}\|_{21}.$$

Proof:

1. Let $\{x_i\}_{i=1}^{n}$ be any normalized λ-unconditional basic sequence in X and $\{a_i\}_{i=1}^{n}$ any sequence of scalars, further denoting by $\{a_{\pi(i)}\}_{i=1}^{n}$ the non-increasing rearrangement of $\{|a_i|\}_{i=1}^{n}$ (where π is the permutation of $\{1, 2, \ldots, n\}$ which effects this rearrangement). Then, for each $1 < k \leq n$,

$$\left\| \frac{\sum_{i=1}^{k} a_{\pi(i)} x_{\pi(i)}}{a_{\pi(k)}} \right\| \geq \frac{1}{\lambda} \left\| \sum_{i=1}^{k} x_{\pi(i)} \right\| \geq \frac{k^{1/2}}{\lambda f(\lambda)}.$$

147

Thus, $\frac{1}{\lambda f(\lambda)} k^{1/2} \|a_{\pi(k)}\| \leq \|\sum_{i=1}^{k} a_{\pi(i)} x_{\pi(i)}\| \leq \lambda \|\sum_{i=1}^{k} a_i x_i\|$, for such k.

Hence $\frac{1}{\lambda^2 f(\lambda)} \sup_{i \leq k \leq n} k^{1/2} a_{\pi(k)} \leq \|\sum_{i=1}^{n} a_i x_i\|$.

For the other inequality, let $\{x_i^*\}_{i=1}^{n}$ be the functionals biorthogonal to the vectors $\{x_i\}_{i=1}^{n}$.

Then for any $A \subset \{1, 2, \ldots, n\}, |A| = \left(\sum_{i \in A} x_i^*\right)\left(\sum_{i \in A} x_i\right) \leq \|\sum_{i \in A} x_i^*\| \|\sum_{i \in A} x_i\| \leq \|\sum_{i \in A} x_i^*\| f(\lambda) |A|^{1/2}$,

so $\|\sum_{i \in A} x_i^*\| \geq \frac{|A|}{f(\lambda)}$. It follows by the above argument, that for any $\{b_i\}_{i=1}^{n}$, if $\{b_{\pi(i)}\}_{i=1}^{n}$ yields the non-increasing rearrangement of $\{|b_i|_{i=1}^{n}\}$, then

$$\frac{1}{\lambda^2 f(\lambda)} \sup_{1 \leq k \leq n} k^{1/2} |b_{\pi(k)}| \leq \|\sum_{i=1}^{n} b_i x_i^*\|.$$

So, given any $\{a_i\}_{i=1}^{n}$, choose $\{b_i\}_{i=1}^{n}$ so that $\|\sum_{i=1}^{n} b_i x_i^*\| = 1$, while $\|\sum_{i=1}^{n} a_i x_i\| = \sum_{i=1}^{n} a_i b_i$.

Then, if $\{|b_{\pi(i)}|\}_{i=1}^{n}$ is the nonincreasing rearrangement of $\{|b_i|\}_{i=1}^{n}$,

$$\begin{aligned} \|\sum_{i=1}^{n} a_i x_i\| &= \sum_{i=1}^{n} a_i b_i = \sum_{i=1}^{n} \left(\frac{a_{\pi(i)}}{i^{1/2}}\right)\left(b_{\pi(i)} i^{1/2}\right) \\ &\leq \left(\sup_{1 \leq i \leq n} |b_{\pi(i)} i^{\frac{1}{2}}|\right) \sum_{i=1}^{n} |\frac{a_{\pi(i)}}{i^{1/2}}| \\ &\leq \lambda^2 f(\lambda) \|\sum_{i=1}^{n} b_i x_i^*\| \sum_{j=1}^{n} \frac{|a_j^*|}{j^{1/2}} \\ &= \lambda^2 f(\lambda) \|\{a_i\}_{i=1}^{n}\|_{21}. \end{aligned}$$

((2) follows likewise, with only notational changes). $\square$

Next, we investigate what the theory of spreading models looks like in an F.D.D. setting.

Definition Ae7:

1. Given a Banach space X, a F.D.D. $\sum_{n} \oplus E_n$ for X is E-invariant under spreading if there exists a Banach space E and operators $T_n : E_n \to E, (n = 1, 2, \ldots)$, such that $\|T_n\| \|T_n^{-1}\| \leq 1 + \frac{1}{2^n}$, and for any $k \leq n_1 < \ldots < n_k$ and any $k \leq m_1 < \ldots < m_k$, and any $x = \sum_{n} x_n \in X$, $|\|\sum_{i=1}^{k} x_{n_i}\| - \|\sum_{i=1}^{k} T_{m_i}^{-1} T_{n_i} x_{n_i}\|| \leq \frac{1}{2^k}$.

2. If $X = \sum \oplus E_n$ is E-invariant under spreading, for each n we define $\Delta(E_n \oplus E_{n+1}) := \{x - T_{n+1}^{-1} T_n x : x \in E_n\}$.

If $\sum_{n} \oplus E_n$ is any F.D.D. for X with $\dim E_n = k < \infty$, for every n, then the compactness of the k-dimensional spaces in the Banach-Mazur distance implies that there exists a k-dimensional space E and integers $n_1 < n_2 < \ldots$ such that $d(E_{n_i}, E) \leq 1 + \frac{1}{2^i}, (i = 1, 2, \ldots)$. If we then apply the Brunel-Sucheston procedure [2], we obtain in this setting:

Theorem Ae8: If $X = \sum_{n} \oplus E_n$ is a F.D.D. with $\dim E_n = k < \infty$ for every n, then there exists a Banach space E and integers $n_1 < n_2 < \ldots$ such that $\sum \oplus E_{n_i}$ is E-invariant under spreading. Moreover, $\sum_{i=k}^{2k} \Delta(E_{n_{2i}} \oplus E_{n_{2i+1}})$ is a 3-U.F.D.D., $(k = 1, 2, \ldots)$.

148

Now we are ready for the first theorem [26] of W.B. Johnson:

Theorem Ae9: There is a function $g : [1,\infty) \to 1],\infty)$ satisfying: if $X = \sum_n \oplus E_n$ is a K-U.F.D.D. and X satisfies property H with function $f(\lambda)$, then

1. If $E = E_n, (n = 1, 2, \ldots)$, then $d_E \leq g(f(K))$.

2. If $\sum_n \oplus E_n$ is E-invariant under spreading, then $d_E \leq g(f(K))$.

Proof:

1. By Proposition Ae6, for every $x = \sum x_n \in X$,

$$(*) \qquad \frac{1}{K^2 f(K)} \|\{\|x_n\|\}_{n=1}^{\infty}\|_{2\infty} \leq \|x\| \leq K^2 f(K) \|\{x_n\}_{n=1}^{\infty}\|_{21}.$$

By a result of Kahane [13], the norms of the spaces $L_p(E)$ are all equivalent on $\mathrm{Rad}(E), (0 < p < \infty)$. Thus the norms of the Lorentz spaces $L_{2\infty}(E)$ and $L_{21}(E)$ are also. Next we observe that for any $\{x_i\}_{i=1}^{n} \subset E$,

$$\left\|\sum_{i=1}^{n} \epsilon_i x_i\right\|_{L^{(E)}_{2\infty}} = 2^{-\frac{n}{2}} \|\{\sum_{i=1}^{n} \epsilon_i x_i\}_{\epsilon \in \{-1,1\}^n}\|_{\ell_{2\infty}^{(E)}}.$$

(We have a similar equality for $L_{21}^{(E)}$). It now follows from $(*)$ that $\mathrm{Rad}_n(E)$ embeds uniformly into X up to a function of $K^2 f(K)$ and the universal constant C of equivalence of the $L_{2\infty}(E)$ and $L_{21}(E)$ norms on $\mathrm{Rad}(E)$. Thus $\mathrm{Rad}(E)$ possesses property H up to a function of $K^2 f(K)$ and C. But then, by Theorem Ac20, $d_E \leq g(f(K))$, for some function g.

2. (We adjust slightly the above argument). Consider the vector space $\hat{E} := \{(x_1, x_2, \ldots) | x_i \in E$, and there exists n such that $x_i = 0$, for $i > n\}$. Define a norm on $\hat{E}$ by: $\|(x_i, x_2, \ldots)\|_{\hat{E}} := \|\sum_n T_n^{-1} x_n\|_X$, where $T_n : E_n \to E$ are the isomorphisms in the definition of $\sum_n \oplus E_n$ being E-spreading. Then $\hat{E} := \sum_n \oplus F_n$ is a 2K-U.F.D.D. Unfortunately, $\hat{E}$ need not be isomorphic to E (or we would be done!). However, we do have:

$$\frac{1}{(2K)^2 f(2K)} \|\{\|x_n\|\}\|_{\ell_{2\infty}} \leq \|\sum x_n\|_{\hat{E}} \leq (2K)^2 f(2K) \|\{\|x_n\|\}\|_{\ell_{21}}.$$

Also, $\|\{\|x_n\|\}\|_{\ell_{2\infty}}\| = \{\|T_n^{-1} x_n\|\}\|_{\ell_{2\infty}}$. This says the the operator $T : \hat{E} \to X$ given by $T(\sum_n x_n) = \sum_n T_n^{-1} x_n$ is a $g(2(2K)^2 f(2K))$-isomorphism on $\mathrm{Rad}_n(E)$. Part (A) can now be invoked. $\square$

Before we prove our next result, let us review the "standard technique" for producing F.D.D.'s in a space X. For any finite-dimensional subspace $E \subset X$ and any $\varepsilon > 0$, there exists a subspace F of X of finite codimension so that $[E, F] \approx_{1+\varepsilon} E \oplus F$. By iterating this argument, we can produce subspaces $F_1 \supset F_2 \supset \ldots$ of finite codimension in X such that

$$\left[\sum_{i=1}^{n} \oplus E_i, F_{n+1}\right] \approx_{1+\frac{1}{2^{n+1}}} \left(\sum_{i=1}^{n} \oplus E_i\right) \oplus F_{n+1}, (n = 1, 2, \ldots).$$

149

Now choose $E_{n+1} \subset F_{n+1}$ with $\dim E_{n+1} < \infty$. Then $\sum_{n} \oplus E_n$ is a F.D.D. for a subspace of X.

We have now laid the foundations for an important result [26] of W.B. Johnson.

Theorem Ae10: Every Banach space X with property H is asymptotically Hilbertian.

Proof: (We proceed by contradiction.) Suppose X has property H, but is not as. Hilbertian. Then, for any $C > 0$, there exists an integer n such that every subspace Y of finite codimension in X contains a subspace E with $\dim E = n$ and $d_E \geq C$. Using the "standard technique" (given above) for producing F.D.D.'s, we can produce a sequence $\{E_k\}_{k=1}^{\infty}$ of n-dimensional subspaces of X with $d_{E_k} \geq 3$, $(k = 1, 2, \ldots)$, such that $\{E_k\}_{k=1}^{\infty}$ is a 2-F.D.D. for $Y := [E_k]_{k=1}^{\infty}$ in X (i.e., $[E_k]_{k=1}^{\infty}$ is the closed linear span of $\{E_1, E_2, \ldots\}$). Theorem Ae8 now implies the existence of a Banach space E and a sequence $n_1 < n_2 < \ldots$ of integers such that $\sum_{i} \oplus E_{n_i}$ is E-invariant under spreading. Then $\sum_{i} \Delta(E_{n_{2i}} \oplus E_{n_{2i+1}})$ is also a 2-F.D.D. and so has a subsequence $\sum_{j} \Delta(E_{n_{2k_j}} \oplus E_{n_{2k_j+1}})$ which is F-invariant under spreading for some space F. Again, by Theorem Ae8, $\sum_{j=m}^{2m} \Delta\left(E_{n_{2k}} \oplus E_{n_{2k+1}}\right)$ is a 3U.F.D.D., so (by Theorem Ae9) $d_F \leq g(f(3))$. It's immediate that $d(E, \Delta(E_{n_2} \oplus E_{n_{2i+1}})) \leq 2$, and so $d(E, F) \leq 3$. Hence $d_E \leq 3g(f(3))$. But $d_E \geq C$, for large C, is a contradiction. $\square$

Finally, W.B. Johnson [26] obtained the following:

Theorem Ae11: Every Banach space X which is as. Hilbertian is reflexive.

Proof: This is immediate from the finite tree property [9], since this shows that for n large enough there exists a subspace Y_n of X of finite co-dimension which is reflexive. Hence, X is reflexive. $\square$

Corollary Ae12: Every weak Hilbert space is reflexive.

We conclude with an observation of P.G. Casazza which says there is no "theory" of non-separable weak Hilbert spaces.

Theorem Ae13: If X is as. Hilbertian, then $X \approx H \oplus Y$, where H is a Hilbert space and Y is a separable Banach space.

Proof: Select finite-codimensional subspaces $Y_1 \supset Y_2 \supset \ldots$ of X such that every n-dimensional subspace of Y_n is K-Euclidean. Now let $Z := [E_n^{\perp}]_{n=1}^{\infty}$ in X^*, where $E_n^{\perp} := \{F \in X^* : F|_{E_n} = 0\}$. Since Z is separable and (by Theorem Ae11) X^* is reflexive, we can apply Lindenstrauss' Theorem [12] to obtain a separable space $Z \subset Y \subset X^*$ such that there exists a projection $P : X^* \to Y$. So $X^* \approx Y \oplus (I - P)X^*$, and $(I - P)X^*$ is a Hilbert space. So $X \approx Y^* \oplus H$. $\square$

Corollary Ae14: If X is a weak Hilbert space, then X is isomorphic to $H \oplus Y$, where Y is a separable weak Hilbert space and H is a (possibly non-separable) Hilbert space.

Notes and Remarks.

1/. We say that a Schauder decomposition $\sum \oplus X_n$ is a Schauder decomposition of copies of X if $\sup_n d(X_n, X) < \infty$. In retrospect, W.B. Johnson's proof of Theorem Ae10 yields:

Theorem Ae15: If $\sum \oplus X_n$ is a Schauder decomposition of copies of X and if $\sum \oplus X_n$ is a weak Hilbert space, than X is isomorphic to a Hilbert space.

2/. There are known examples of spaces which are as. Hilbertian but fail property H (and hence are not weak Hilbert spaces). One such is due to W.B. Johnson [10], who constructs a Banach space of form $X = (\sum \oplus \ell_{p_n}^{k_n})_{\ell_2}$ such that every subspace of every quotient space of X has form $(\sum \oplus E_n)_{\ell_2}$, where $\dim E_n < \infty$, $(n = 1, 2, \dots)$.

It is easily checked that this space is as. Hilbertian but fails property H.

3/. It follows from Theorem Ae15 that any space with a subsymmetric basis or with a symmetric basis is a weak Hilbert space iff it is isomorphic to a Hilbert space. So $S(T^{(2)})$ is not a weak Hilbert space, despite having the property (see Proposition Ab10) that, for any subspace E, $d_E = o(\log_m \dim E)$.

Af: Open Problems.

Here we discuss some of the important open problems in the theory of weak Hilbert spaces. To put these into perspective, we will simultaneously develop some new theory related to these questions.

The first and foremost problem in the theory of weak Hilbert spaces is to find more examples of such spaces. Let us review the few examples available: the same techniques which worked for $T^{(2)}$ suffice to show that $T_\theta^{(2)}$ is a weak Hilbert space, for all $0 < \theta < \frac{1}{2}$, (where $T_\theta^{(2)}$ is the 2-convexification of T_θ, as defined in section X.A.). It follows that for any such $\theta_1, \theta_2, \dots, \theta_n$, and $\theta_1', \theta_2', \dots, \theta_m'$, every subspace of every quotient space of the space:

$$\sum_{i=1}^{n} \oplus T_{\theta_i}^{(2)} \oplus \sum_{j=1}^{m} \left(T_{\theta_j'}^{(2)} \right)^* \oplus \ell_2$$

is a weak Hilbert space (and, by the argument of W.B. Johnson [10], has a basis). At this point, these are the only examples known, so we formulate:

Problem Af2: Find more examples of weak Hilbert spaces.

Also, since each example we know has a basis, in light of G. Pisier's proof that every weak Hilbert space has the approximation property, we also ask:

Problem Af3: Does every weak Hilbert space have a basis?

Now we review what is known in this direction. P.G. Casazza [3] has shown:

Theorem Af4: If X^* is separable and has the approximation property, then there exists a

151

subspace Y of X such that both Y and $X \oplus Y$ have F.D.D.'s.

Since weak Hilbert spaces are stable with respect to taking subspaces and finite direct sums, we have:

Theorem Af5: Every separable weak Hilbert space is complemented in a weak Hilbert space with a F.D.D.

W.B. Johnson's proof [10] shows in particular:

Theorem Af6: Let X be a Banach space, and suppose there is a family of subspaces $E_1 \supset E_2 \supset \cdots$ of X such that:

1. codim $E_n < \infty$, for every n, and
2. there is a constant β such that every subspace F of E_n with $\dim F \leq 5^{\operatorname{codim} E_n}$ satisfies $d_F \leq \beta$.

Then every subspace of every quotient space of X has a F.D.D.

As an immediate corollary, we have:

Theorem Af7: If X is a weak Hilbert space and X has an unconditional finite dimensional decomposition ("U.F.D.D."), then X is isomorphic to some $X_1 \oplus X_2$, where every subspace of every quotient space of X_i has a F.D.D., (for $i = 1, 2$).

So it becomes natural to ask:

Problem Af8: Does every weak Hilbert space have a U.F.D.D.?

In fact, the following is unknown:

Problem Aaf9:

(a). Does every weak Hilbert space have an unconditional basis?

(b). Does every subspace of $T^{(2)}$ have an unconditional basis?

This leads to yet another question:

Problem Af10: Classify all weak Hilbert spaces (or classify those with unconditional bases).

Note that at this point the only weak Hilbert spaces known are subspaces of quotients of variations of $(T^{(2)})^*$ and $(\sum \oplus E_n)_{\ell_2}$. Recall that P.G. Casazza, W.B. Johnson, and L. Tzafriri argued in [4] (see Theorem III.6, these notes), using the fact that subspaces of quotient spaces of $T^{(2)}$ have bases, that:

Theorem Af11: Every subspace of every quotient space of $T^{(2)}$ is a $\left(T^{(2)} \right)$-Tsirelson-sum of finite dimensional spaces.

It is likely that there exist weak Hilbert spaces which fail to have unconditional bases. In this direction we ask:

Problem Af12: Does every weak Hilbert space embed into a weak Hilbert space with an unconditional basis?

We know (by Theorem Ad11) that every weak Hilbert space X has the uniform approximation property, i.e., that there exists $K > 0$ and for each n there exists $k(n)$ such that for every finite dimensional subspace E of X there is a finite rank operator $T : X \rightarrow X$ such that $T|_E = Id|_E$, $\|T\| \leq K$, and rank $T \leq k(\dim E)$. So we ask:

Problem Af13:

1. Does there exist (for every weak Hilbert space X) a constant c such that $k(n) \leq cn$?
2. Does the existence of such a c for a Banach space X imply that X is a weak Hilbert space?

Despite all these "strong" properties for $T^{(2)}$ and weak Hilbert spaces, we cannot resist formulating a problem in the opposite direction. It is conceivable that a variation of the $T^{(2)}$ construction could be made in a "conditional" manner. That is, perhaps we can manage to construct some variant X of $T^{(2)}$ so that for each finite-dimensional subspace E of X, d_E is proportional to the unconditional basis constant of E. If this is in the cards, we would have a very strong counter-example to the famous old:

Problem: Does every Banach space contain unconditional basic sequences?

In particular, we ask:

Problem Af14: Does there exist a weak Hilbert space X such that every subspace of every quotient space of X has a basis, but no subspace of any quotient space of X has an unconditional basic sequence?

Related to properties H and H_2 (see Def. Ad1), we ask:

Problem Af15:

1. Does property H imply property H_2?
2. Does property H for a space X imply X is a weak Hilbert space?

Note that the weak Hilbert space property and the as. Hilbertian property are stable under quotients, subspaces, and the taking of duals. Properties H and H_2 are "between" these, so we ask:

Problem Af16: Are properties H and H_2 stable under the formation of quotients, and duals?

Recalling "Maurey's extension property" (Theorem Aa6), we see that any space with this property is weak type 2. It is unknown if such a space is type 2, but we ask:

Problem Af17: Does $(T^{(2)})^*$ have the "Maurey extension property"?

Also, recall Theorem Aa4: any operator from a type 2 space to a cotype 2 space factors through a Hilbert space.

This suggests:

Problem Af18: Does every operator from a weak type 2 space to a weak cotype 2 space factor through a weak Hilbert space?

Recall that the converse of this factorization theorem is still unknown, so we ask further:

Problem Af19: Does every operator from $S(T^{(2)})$ to $(T^{(2)})^*$ factor through a Hilbert space?

Also, it is unknown if there is a Banach space with a symmetric basis for which every subspace has a basis. The most likely candidate for such a space seems to us to be $S(T^{(2)})$. We ask:

Problem Af20: Does every subspace of $S(T^{(2)})$ have a basis?

There is a whole cycle of ideas related to the Maurey properties, many of which can probably be sharpened or improved. We have the "Maurey extension property" (or "M.E.P.") and the "Maurey projection property" (or "M.P.P.") from Theorem Aa8. The definition of weak type 2 space gives a "weak Maurey extension property" (or "W.M.E.P."). We can similarly define:

Definition Af21: A Banach space X has the weak Maurey projection property ("W.M.P. P.") if for every $0 < \delta < 1$ there is a $C_\delta > 1$ and a $K > 1$ such that for any subspace E of X with dim $E = n$ and $d_E \leq C_\delta$, there is a subspace F of E with dim $F \geq \delta n$ and a projection $P : X \to F$ with $\|P\| \leq K$.

We know that:

1. A type 2 space has M.E.P. (This is Maurey's Theorem Aa8, but the converse is unknown).
2. A space with M.E.P. is wT_2 (but the converse is unknown).

For our next theorem, we require a remarkable result of V. Milman [19]:

Theorem Af22: There is a function $c(\delta), (0 < \delta < 1)$, such that for every n and every Banach space X of dimension n, there exists a subspace E of a quotient space of X such that dim $E \geq \delta n$ and $d_E \leq c(\delta)$.

Now we can prove:

Theorem Af23: For a Banach space X, the following are equivalent:

1. X is wT_2.
2. Every quotient space of X is wT_2 (uniformly).

3. Every quotient space of X has W.M.P.P. (uniformly).

Proof: (2) implies (3) is immediate. To see that (1) iff (2), note that X being wT_2 implies X is K-convex, and so every quotient Y of X is (uniformly) K-convex. Since X^* is wC_2 and Y^* embeds into X^*, Y^* is wC_2, whence Y is wT_2.

(3) implies (1): Note that (3) implies X is K-convex. So (by Theorem Ac16) we need to show that X^* is wC_2. Let E be a finite dimensional subspace of X^*. Then E^* is a quotient of X. By Milman's Theorem, there exists a quotient F of E^* with $\dim F \geq c \dim E^*$, and a subspace G of F with $\dim G \geq f(c) \dim E^*$. Since F is a quotient of X, (by (3)) G is complemented (uniformly) in F. Thus G is a quotient of E^* and so G^* embeds (uniformly) into E. But $d_{G^*} \leq f(c)$. So X^* is wC_2, and hence X is wT_2. $\square$

Continuing the discussion following Definition Af21, we can also observe:

3. wT_2 implies W.M.P.P. (but the converse is unknown).

4. If X is wC_2, then X has W.M.P.P. iff X is wT_2 iff X is a weak Hilbert space.

5. M.E.P. implies M.P.P. (but the converse is unknown).

6. M.E.P. implies wT_2 and M.P.P. (but the partial converses are unknown).

7. It is known that $\ell_2(X)$ has M.E.P. iff it has W.M.E.P. iff X is type 2, (but it is unknown whether $\ell_2(X)$ having M.P.P. or W.M.P.P. implies X is of type 2).

Related to all this, we give:

Definition Af24: A space X is finite dimensionally norming ("F. D.N."), if for every $0 < \delta < 1$, there is a $C_\delta > 1$ such that for every subspace E of X with $\dim E = n$, there is a subspace F of X^* with $\dim F \geq \delta n$ such that E is "C_δ-norming" over F i.e.,

$$\frac{1}{C_\delta}\|f\| \leq \sup\{|f(x)| : \|x\| = 1, \ x \in E\}, \qquad \text{for all } f \in F.$$

Continuing:

8. It is easily seen that every weak Hilbert space is F.D.N. (although the converse is unknown).

9. It is also easily seen that each F.D.N. space has M.P.P. (although here too the converse is open).

Recall also:

Definition Af25: A Banach space X has the Orlicz property ("O.P.") if for every unconditionally summable series $\sum_n x_n$ in X, we have $\sum_n \|x_n\|^2 < \infty$.

In [26], it is observed that X has cotype 2 iff $(\sum \oplus X)_{\ell_1}$ has O.P. The relationship between wC_2 and O.P. is unknown (although clearly O.P. with wT_2 implies property H).

Definition Af26: A Banach space X is well normed ("W.N.") if for every $1 < \delta$, there is a $C_\delta > 1$ such that for every subspace E of X there is a subspace F of X^* with $\dim F \leq \delta \dim E$

and F is C_δ-norming over E.

It is unknown how the property W.N. fits into these other properties, but if it could be shown that every weak Hilbert space is W.N., then the constructions of W.B. Johnson [10] could be done in an abstract weak Hilbert space to answer many of the questions posed earlier. Also, volume ratios fit into this cycle somewhere. Under some circumstances they produce good Euclidean subspaces and in others they provide projections onto these subspaces. However, the symmetrical nature of volume ratios should allow one, given a finite dimensional subspace E of X, to produce projections onto subspaces F of X with $E \subset F$ and $\dim F \leq \delta \dim E$, for some fixed $\delta > 1$. None of this seems to be understood yet, but all of these concepts are closely aligned with the theory.

There are more than a few other problems of interest related to weak Hilbert spaces, but space considerations force us to stop here.

Notes and Remarks:

1/. Related to problem Af20, P.G. Casazza has shown:

Theorem Af27: If X is a subspace of $S(T^{(2)})$, then either:

1. $S(T^{(2)})$ embeds complementably in X (and hence X has a basis iff X has the approximation property (see [3])), or

2. X is as. Hilbertian.

¿From this, it seems likely that the answer to Problem Af20 is "yes".

2/. It can be shown, using a result of J. Bourgain and L. Tzafriri [1] that type 2 plus property H implies property H_2. Actually, weak type 2 should suffice, and we have a suspicion that these two properties are more than likely equivalent.

Weak Hilbert Spaces: a bibliography

[1] Bourgain, J. & Tzafriri, L.: Invertibility of "large" submatrices with applications to the geometry of Banach spaces and harmonic analysis, (pre-print).

[2] Brunel, A. and Sucheston, L.: On J-convexity and some ergodic super-properties of Banach spaces, Trans. A.M.S. 204 (1975), pp. 75-90.

[3] Casazza, P.G.: A remark on the approximation property for Banach spaces (pre-print).

[4] Casazza, P.G., Johnson, W.B., & Tzafriri, L.: On Tsirelson's space, Israel J. Math. 47 (1984), pp. 81-98.

[5] Dilworth, S. & Szarek, S.J.: The cotype constant and an almost Euclidean decomposition for finite-dimensional normed spaces, Israel Journal Math. 52 (1985), pp. 82-96.

[6] Figiel, T., Lindenstrauss, J., & Milman, V.: The dimension of almost spherical sections of convex bodies, Acta Math. 139 (1977), pp. 53-94.

[7] Figiel, T., & Tomczak-Jaegermann, N.: Projections onto Hilbertian subspaces, Israel Journal Maths. 33 (1979), pp. 155-171.

[8] James, R.C.: Nonreflexive spaces of type 2, Israel Journal Math. 30 (1978), pp. 1-13.

[9] James, R.C.: Some self-dual properties of normed linear spaces, Annals of Math. Studies 69 (1972).

[10] Johnson, W.B.: Banach spaces all of whose subspaces have the approximation property, Séminaire d'Analyse Fonctionnelle 79/80, Ecole Polytechnique Palaiseau, Exposé No. 16.

[11] Lewis, D.: Ellipsoids defined by Banach ideal norms, Mathematika 26 (1979), pp. 18-29.

[12] Lindenstrauss, J.: On non-separable reflexive Banach spaces, Bull. A.M.S. 72 (1966), pp. 967-970.

[13] Lindenstrauss, J. and Tzafriri, L.: Classical Banach Spaces II: Function Spaces, Springer-Verlag, 1979.

[14] Lindenstrauss, J. & Tzafriri, L.: The complemented subspaces problem, Israel Journal Math. 9 (1971), pp. 263-269.

[15] Maurey, B.: Un théorème de prolgement, C.R. Acad. Sci. Paris, A279 (1974), pp. 329-332.

[16] Maurey, B. & Pisier, G.: Séries de variables aléatoires vectorielles indépendantes et propriétes gésmetriques d espaces de Banach, Studia Math. 58 (1976), pp. 45-90.

[17] Milman, V. & Pisier, G.: Banach spaces with a weak cotype 2 property, Israel Journal Math. 54 (1986), pp. 139-158.

[18] Milman, V. & Schechman, G.: Asymptotic theory of finite-dimensional normed spaces, Springer-Verlag Lecture Notes in Math. vol. No. 1200 (1986).

[19] Milman, V.: Random subspaces of proportional dimension of finite-dimensional normed spaces: approach through the isoperimetric inequality, Proceedings of the Missouri Con-

ference 1984, Springer Lecture Notes in Math vol. No. 1166, pp. 106-115.

[20] Mascioni, V. and Matter, U.: On the weak cotype and weak type of Banach spaces (pre-print).

[21] Pajor, A.: Quotient volumique et espaces de Banach de type 2 faible, Israel Journal Math. (1987), (pre-print).

[22] Pietsch, A.: *Operator Ideals*, North-Holland (1978).

[23] Pisier G.: Un théorème de factorisation pour les opérateurs liniéaires entre espaces de Banach, Ann. Ecole Normale Sup. 13 (1980), pp. 23-43.

[24] Pisier, G.: Factorization of linear operators and geometry of Banach spaces, C.B.M.S. No. 60, A.M.S. (1985).

[25] Pisier, G.: Holomorphic semi-groups and the geometry of Banach spaces, Annals of Math. 115 (1982), pp. 375-392.

[26] Pisier, G.: Weak Hilbert spaces, (pre-print).

[27] Szarek, S.J.: On Kashin's almost orthogonal decomposition of ℓ_1^n, Bull. Acad. Polon. Sci. 26, pp. 691-694.

[28] Szarek, S.J. and Tomczak-Jaegermann, N.: On nearly Euclidean decompositions for some classes of Banach spaces, Comp. Math. 40 (1980), pp. 367-385.

[29] Tomczak-Jaegermann, N.: *Finite-dimensional operator ideals and Banach-Mazur distances*, (pre-print).

AN ALGORITHM FOR COMPUTING THE
TSIRELSON'S SPACE NORM

Johnnie W. Baker
Kent State University

Oberta A. Slotterbeck
Hiram College

Richard Aron
Kent State University
and
Trinity College

I. INTRODUCTION.

Two programs, NORM and TRACE, are included in this appendix. NORM was created to calculate the Tsirelson's space (or T) norm for vectors with finitely many nonzero terms. The space T used here is the dual of Tsirelson's original space as defined by Figiel and Johnson. The norm used will be explained in Section III. It is easy to adjust the program to calculate other definitions of the norm. The program TRACE includes the code for NORM, but allows the user the choice of tracing intermediate results as they are generated. It is included here as we believe these intermediate calculations may prove interesting to the researcher. When some of the tracing features are being used, TRACE runs slower than NORM.

Both programs were created and tested on a VAX 11/780 running under UNIX (version 4.2 UCB). The programs are in standard Pascal and should run with little or no modifications on any computer supporting standard PASCAL. A few changes may be necessary to use UCSD Pascal, typically found on microcomputers. The algorithm used can be implemented in other languages such as FORTRAN, C, or BASIC if desired.

We have tried to include features in the programs that we believe most mathematicians would want. However, it is impossible to either include every desirable feature or to anticipate the needs of every user. If you use these programs extensively, you will undoubtedly want some features to be changed or added. If you are unfamiliar with programming, your computer center staff should be able to provide this service.

To allow the two programs to be easily available to potential users, the Department of Mathematical Sciences at Kent State University has agreed to prepare distribution tapes containing the Pascal source code for the two programs listed in Section X and Section XI. A copy can also be obtained by electronic mail if you are on a network that can access CSNET. A nominal fee may be charged to cover expenses. The source code was prepared for a VAX/UNIX (Berkeley) environment, but it should be relatively easy to transport it to other Pascal environments. For further information, please write to Johnnie W. Baker, Department of Mathematical Sciences,

Kent State University, Kent, Ohio 44242 (or CSNET address: jbaker@kentvax).

II. BASIC INSTRUCTIONS.

The programs are designed to be run interactively by a user at a terminal or on a personal computer. In this appendix, printout is displayed between dashed lines. The user's input always appears after the input prompt ">>". All other printout is generated by the program. Discussions about the printout follow the display.

A typical session for the NORM program is included below:

```
-------------------------------------------------------------------------

Do you want instructions? (y/n) >> y


Enter vector components, separated by one or more spaces
following the ''Enter vector >> '' prompt. For example:
Enter vector >> 4 4 4 7 7 7 7
If you can not enter all of the components on one line,
continue entering the components on the next line, either by
running off the end and continuing, or by hitting the return
key immediately after a space has been entered.
DO NOT enter a space after the last vector component.
When you are instructed to enter (y/n), enter ''y'' for yes
and ''n'' for no. DO NOT enter a space before the ''y'' or ''n''.


Enter vector >> 4 4 4 7 7 7 7


Enter maximum number of (additional) norm iterations >> 6
          0-norm =    7.00
          1-norm =   10.50
          2-norm =   10.50
The norms stabilized after   2 iterations
Do you wish to process another vector? (y/n) >> y


Enter vector >> 7 7 7 7 4 4 4


Enter maximum number of (additional) norm iterations >> 1
          0-norm =    7.00
          1-norm =    7.50
```

160

```
Do you wish to continue processing the same vector? (y/n) >> y
Enter maximum number of (additional) norm iterations >> 3
         2-norm =    7.50
The norms stabilized after   2 iterations
Do you wish to process another vector? (y/n) >> y


Enter vector >> 1 2 3 4 5 6 7 8 9 0 1 2 3 4 5 6 7 8 9 0 1 2 3 4 5
6 7 8 9 0 1 2 3 4 5 6 7 8 9 0


Enter maximum number of (additional) norm iterations >> 4
         0-norm =    9.00
         1-norm =   52.50
         2-norm =   53.75
         3-norm =   53.75
The norms stabilized after   3 iterations
Do you wish to process another vector? (y/n) >> n
```

The user should answer the questions by answering the preferred response, followed by a (carriage) return. As Pascal's input is rather sensitive, a few points should be noted. Do not enter a space prior to entering a response of "y" for "yes" or "n" for "no" (without the quotes). To enter a vector that is longer than one line, the user can continue to enter the components and let the cursor run off the right side of the screen and wrap around to the left. Be sure your terminal is set to wrap and that you do not enter a return if this approach is taken. Alternately, the user can enter a space or spaces after the last number on the right, followed by a return. Integers or numbers in decimal form may be vector components. When a user asks for instructions, a summary of the above information is printed on the screen. (See session above.)

The norm calculations for TRACE use the same code as NORM. The primary difference between the two programs is that TRACE allows the user to trace intermediate calculations. Therefore, the code for TRACE is longer than the code for NORM. The meaning of the printed values will be discussed later in this appendix. The displays are different for vectors of length less than or equal to 13 and those of length 14 or more (because of the physical length of a display line on the terminal). As the output from TRACE can be quite extensive on vectors of length greater than 13, if much tracing is to be done the user will probably wish to have the program output to a file. A person familiar with the computer being used will be able to provide instructions on how to do this.

A typical session for TRACE is given below. Details of TRACE are covered in Section IX.

```
Do you want instructions? (y/n) >> n
Do you want a trace of the calculations printed? (y/n) >> y
Do you wish to have intermediate results printed? (y/n) >> n
Do you wish to have the newnorm update results printed? (y/n) >> n


Enter vector >> 5 4 3 2 1
  5.00  4.00  3.00  2.00  1.00
The vector has length    5
Enter maximum number of (additional) norm iterations >> 3
The current total number of norm iterations permitted is    3
Now processing vector for level =  0


VALUES OF SUP-NORMS OF SUBVECTORS
low=(row nr)     high=(column nr  >
4.00      4.00      4.00      4.00
          3.00      3.00      3.00
                    2.00      2.00
                              1.00
                              5.00
The vector norm value for this level is          0-norm =    5.00
Now processing vector for level =  1
normsums[1,.] = oldnorms[.] for parts = 1


VALUES OF NORMSUMS MATRIX FOR PARTS =   2  WITH LEVEL =    1
low=(row nr)     high=(column nr) >
0.00      0.00      0.00      0.00
          0.00      5.00      5.00
                    0.00      3.00
                              0.00


UPDATE OF MATRIX NEWNORMS WITH LEVEL =    1
low=(row nr)     high=(column nr) >
4.00      4.00      4.00      4.00
          3.00      3.00      3.00
```

162

```
        2.00      2.00

                  1.00

                  5.00
```

```
The vector norm value for this level is          1-norm =    5.00
The norms stabilized after    1 iterations
```

III. PRELIMINARIES TO THE ALGORITHM.

Let $\mathbb{R}^{(N)}$ denote the space of all sequences with only finitely many nonzero terms. If Y is in $\mathbb{R}^{(N)}$, then Y has the form

$$Y = (Y_1, Y_2, \cdots, Y_n, 0, 0, \cdots)$$

If lo and hi are integers with

$$0 \leq \text{lo} < \text{hi} \leq n$$

then we denote by (lo,hi] the subvector

$$(0, 0, \cdots, 0, Y_{\text{lo}+1}, Y_{\text{lo}+2}, \cdots, Y_{\text{hi}})$$

of Y. The term "subvector" will only be used to refer to a block of consecutive components from the specified vector.

To calculate $\|Y\|_m$, we will first calculate and store $\|W\|_{m-1}$ for all subvectors W of Y. At the present, we assume that these values are all stored in an $n \times (n-1)$ matrix called OLDNORMS, indexed by $(i,j), 1 \leq i \leq n, 2 \leq j \leq n$ (see example below). The norm $\|(\text{lo}, \text{hi}]\|_{m-1}$ is stored at position (lo,hi) in this matrix. The value $\|Y\|_{m-1}$ is stored at (n,n). For example, if $Y = (8, 4, 6, 2)$, and $m - 1$, then the following information is stored in OLDNORMS:

	2	3	4
1	4	6	6
2		6	6
3			2
4			8

Note that approximately half of the storage locations are used. A more efficient storage is actually used. However, it is convenient to visualize storage in the above form in the algorithm description. The storage scheme for another matrix NEWNORMS is the same as for OLDNORMS.

The definition of the norm for Tsirelson's space which is used here is the one described below: A partition P of $1, 2, \ldots n$ is a set

$$P = \{p_1, p_2, \cdots, p_{k+1}\}$$

of integers with

$$k \leq p_1 < p_2 < \cdots < p_{k+1} \leq n.$$

The m-norm is defined inductively by

$$\|Y\|_0 = \max_i |Y_i|$$

and for $m \geq 1$

$$\|Y\|_m = \max \begin{cases} \|Y\|_{m-1} \\ \frac{1}{2}\max_P \sum_{j=1}^{k} \|(p_j, p_{j+1})\|_{m-1} \end{cases}$$

Then the norm on T is defined by

$$\|Y\| = \lim_{m \to \infty} \|Y\|_m$$

Tsirelson's space T is the completion of $\mathbb{R}^{(N)}$ with this norm. The algorithm presented here works only on the subspace $\mathbb{R}^{(N)}$ of the Tsirelson's space T.

In the definition of $\|Y\|_m$ for $m \geq 1$, k is the number of subvectors for partition P. (The k is called PARTS in the actual algorithm description in Section IV.) The amount of work required for the calculation of $\|Y\|$ can be reduced by saving the values

$$Y(k, \text{ lo, hi}) = \max_P \sum_{j=1}^{k} \|(p_j, p_{j+1})\|_{m-1}$$

for successive values of $k(k = 1, 2, \cdots, \lfloor n/2 - 1 \rfloor)$. Here $0 \leq$ lo $<$ hi $\leq n$ and $P_k = \{p_1, p_2, \cdots, p_{k+1}\}$ ranges over all subsets of integers satisfying

$$\max \{k, \text{ lo }\} \leq p_1 < p_2 \cdots < p_{k+1} \leq \text{ hi}.$$

Also, $\lfloor \cdot \rfloor$ denotes the greatest integer function.

It is convenient to assume these values will be stored in the three dimensional $(\lfloor n/2 \rfloor - 1) \times n \times (n-1)$ matrix NORMSUMS with $Y(k, \text{ lo, hi})$ stored at location NORMSUMS $(k, \text{ lo, hi})$. For each fixed k, the values $Y(k, \text{ lo, hi})$ are stored in the $n \times (n-1)$ submatrix NORMSUMS $(k, ., .)$ in the same manner as used for NEWNORMS earlier in this section. These values can then be used to calculate the corresponding values for the $n \times (n-1)$ submatrix NORMSUMS $(k, ., .)$, as indicated in step 13 of the algorithm.

Again, at most half the storage locations are used. A more efficient storage scheme will be introduced after the algorithm is described.

IV. THE ALGORITHM.

In describing the algorithm, we refer to the different procedures in the code in section X. An outline of the code structure may prove helpful. Indented procedures below are called by the preceding procedure which is less indented:

```
program Norm(input,output);

    procedure Instructions: Provides instructions to the user.

    procedure EnterVector: Handles input of the vector.

    procedure ProcessVector: Calculates the m-norms of the current vector
     entered and reports the value of each m-norm as
     it is calculated. If the m-norm values stabilize,
     terminate.

       procedure NextLevel: Calculates the next level of m-norms for
                 the subvectors of the vector entered.

         procedure SumsOfNorms: Calculates values for appropriate sums of
norms of subvectors and stores the results
in NORMSUMS matrix.

         procedure UpdateNorm: Updates the current calculation of the norms
  of subvectors using values in the NORMSUMS
  matrix. Results are stored in the NEWNORMS
  matrix.

End of NORM.
```

The following is a step-by-step outline of the control flow during a run of the program. The name of the procedure containing each step is stated prior to the step. Variables and named constants in the code are introduced as needed and are given in capitals.

1. (EnterVector) Prompt for vector, read it, and set DIM = length of vector

2. (Norm) Set LEVEL to 0.

3. (ProcessVector) Prompt user for highest level M permitted for M-norm and assign this value to TOPLEVEL.

4. (ProcessVector) Set STABILIZED to FALSE.

5. (ProcessVector) {Begin loop to calculate M-norm for each level M successively, $M = 0, 1, \ldots$, TOPLEVEL.} While LEVEL $\leq$ TOPLEVEL and STABILIZED is false, repeat

steps (6) - (26).

6. (ProcessVector) Set STABILIZED to TRUE.

7. (NextLevel) {Initialize NEWNORMS matrix for zero level.} If LEVEL is zero, execute steps (8) - (10), else, go to (11).

8. (NextLevel) Calculate the sup-norm of the subvector from position (LO+1) to position HI and store in the array location NEWNORMS[LO,HI] for all integers LO and HI with $0 \leq$ LO < HI $\leq$ DIM.

9. (NextLevel) {Store the 0-norm of original vector.} Set NEWNORMS[DIM,DIM] to sup-norm of the original vector.

10. (NextLevel) Copy values in NEWNORMS into OLDNORMS.

11. (NextLevel) {Begin calculations for the three dimensional matrix NORMSUMS for this level.} If LEVEL > 0, execute steps (12) - (25) , else go to (26).

12. (NextLevel) Copy the values of the two dimensional array NEWNORMS into the two dimensional submatrix NORMSUMS [1, . , .].

13. (SumsOfNorms) {Next, start calculating values for the two dimensional submatrix NORM-SUMS [PARTS+1, . , .] (i.e. with PARTS fixed) from the two dimensional submatrix NORMSUMS [PARTS, . , .]. For PARTS + 1 starting at 2 until $\lfloor$ DIM/2 $\rfloor$ do steps (14) - (24). $\lfloor$ DIM/2 $\rfloor$ denotes the greatest integer of DIM/2.

14. (SumsOfNorms) For LO starting at PARTS+1 until DIM-1 do steps (15) - (16).

15. (SumsOfNorms) For HI starting at (PARTS+1) + LO until DIM do step (16).

16. (SumsOfNorms) Set NORMSUMS[PARTS+1,LO,HI] to the maximum of the values OLD-NORMS[LO,MID] + NORMSUMS[PARTS,MID,HI] for all integers MID satisfying LO + $1 \leq$ MID $\leq$ HI - PARTS.

17. (UpdateNorm) {Steps (17) - (24) updates the NEWNORMS matrix from m-norms to (m+1)-norms, using the values calculated for the two-dimensional submatrix NORMSUMS [PARTS+1, . , .].} For LO starting at 1 until DIM-1 do block (18) - (23).

18. (UpdateNorm) For HI starting at LO+1 until DIM do steps (19) - (23).

19. (UpdateNorm) Initialize X to zero.

20. (UpdateNorm) {Calculate ASSIGNMENT A value for X.} If PARTS + $1 \leq$ LO and (PARTS + 1) + LO $\leq$ HI, let X be NORMSUMS[PARTS+1,LO,HI].

166

21. (UpdateNorm) {Calculate ASSIGNMENT B value for X.} If LO < PARTS + 1 and $2(PARTS+1) \leq HI$, let X be NORMSUMS[PARTS+1,PARTS+1,HI].

22. (UpdateNorm) {Update norm of vector (LO,HI] using contributions from NORMSUMS[parts+ . , .].} Replace NEWNORMS[LO,HI] with the larger of its present value and X/2.

23. (UpdateNorm) Set STABILIZED to FALSE if the value of NEWNORMS[LO,HI] was increased in step (22).

24. (UpdateNorm) {Update the norm of the original vector for this level.} Set NEWNORMS[DIM,DI to be the larger of its present value and NEWNORM[1,DIM].

25. (NextLevel) Copy the values of the two dimensional array NEWNORMS into the two dimensional array OLDNORMS.

26. (ProcessVector) Set LEVEL to LEVEL + 1 and return to (5).

V. SPACE REQUIREMENTS.

As mentioned earlier, the storage scheme we presented for NEWNORMS and OLDNORMS is inefficient. Instead of using two-dimensional arrays for NEWNORMS and OLDNORMS, we use one-dimensional arrays. The norm information for the subvector (LO,HI] is stored at location $(LO-1)N - LO(LO+1)/2 + HI$ in the linear arrays. For the vector $X = (x_1, x_2, \cdots, x_n)$, this leads to the following storage locations for the subvectors of X :

167

NUMBER OF SUBVECTORS

ARRAY INDEX	SUBVECTOR	MATRIX INDEX	NUMBER OF VECTORS
1	$0, x_2)$	$(1,2)$	
2	$(0, x_2, x_3)$	$(1,3)$	$n-1$
.			
.			
.			
$n-1$	$(0, x_2, x_3, \ldots, x_n)$	$(1, n)$	
n	$(0, 0, x_3)$	$(2,3)$	
$n+1$	$(0, 0, x_3, x_4)$	$(2,4)$	$n-2$
.			
.			
.			
$2n-3$	$(0, 0, x_3, \ldots, x_n$	$(2, n)$	
$2n-2$	$(0, 0, 0, x_4)$	$(3,4)$	
$2n-1$	$(0, 0, 0, x_4, x_5)$	$(3,5)$	$n-3$
.			
.			
.			
$3n-6$	$(0, 0, 0, x_4, \ldots, x_n)$	$(3, n)$	
.			
.			
.			
$n(n-1)/2$	$(0, \ldots, 0, x_n)$	$(n-1, n)$	
$n(n-1)/2+1$	$(x_1, x_2, \ldots, x_n)$	(n, n)	

The storage of the norm of the vector $(x_1, x_2, \cdots, x_n)$ was at (n, n) in the matrix scheme. Under the new approach, we store it at the next location after the storage location for $(0, 0, \cdots, x)$, i.e. at index $n(n-1)/2 + 1$.

The storage of NORMSUMS is also inefficient. For a fixed index PARTS, NORMSUMS[PARTS,LO,HI] is stored in a linear array with

$$\text{INDEX} = (\text{LO-1})\text{N} - \text{LO}(\text{LO+1})/2 + \text{HI}.$$

Thus, NORMSUMS can be regarded as a two-dimensional array and (PARTS,LO,HI) can be stored at NORMSUMS[PARTS,INDEX]. However, one additional space saving scheme is used. After NORMSUMS[PARTS, .] is calculated, it is needed only in the calculation of NORM-SUMS[PARTS+1, .]. After this calculation is completed, NORMSUMS[PARTS, .] can have its storage space used again. As a result, we need only two values for the first component in the index for NORMSUMS. This is accomplished by replacing NORMSUMS[PARTS,INDEX] with NORMSUMS[PARTS mod 2, INDEX].

The total number of storage locations in each array NEWNORMS and OLDNORMS is $n(n-1)/2+1$. The number of storage locations for NORMSUMS is $2[n(n-1)/2+1] = n(n-1)/2$.

168

Also, n locations are required to store the vector X entered by the user. Consequently, the total storage for arrays is $2n^2 - n + 4$. Thus, the storage required is $O(n^2)$, where n is the maximum length that is permitted for the vector entered by the user.

VI. SOME SWITCHES IN THE PROGRAM.

The variable STABILIZED is a switch. It has an initial value of TRUE for each norm level calculated (see step 6). As the norm values for subvectors are updated in step (22), the value of STABILIZED is changed to FALSE if the norm of any subvector is increased. If STABILIZED is still TRUE after the calculations of the NEWNORMS matrix, then the NEWNORMS and OLDNORMS matrices are identical and the calculation for the next level will also be identical. As a result, if STABILIZED is TRUE after the calculation of a given level, the value of all m-norms for all m greater than this level for each subvector will be the same as they are for this level. The norm value for this level for the original vector entered by the user can be returned as the Tsirelson's space norm of the vector.

The variable SUMSGROW, found in procedure SumsOfNorms, is another switch. It has an initial value of FALSE for each pass through the outer "parts" loop. If a larger value is found for the two-dimensional submatrix NORMSUMS[(PARTS+1)mod 2, . , .], i.e. with PARTS fixed, than for the corresponding position in NORMSUMS[PARTS mod 2, . , .] then SUMSGROW has its value changed to TRUE. Otherwise, these two-dimensional submatrices of NORMSUMS are equal and no values in the matrix NEWNORMS will be changed by the procedure UpdateNorms. The calculations for NORMSUMS[(PARTS+2)mod 2, . , .] will be the same as for NORMSUMS[(PARTS+1)mod 2, . , .]. Consequently NORMSUMS[k mod 2, . , .] need not be calculated for $k \geq$ PARTS + 1 and the calculation of the norm for this level is complete.

VII. TIMINGS.

Using the definition of the Tsirelson's space norm, one can develop a more natural algorithm than the preceding one using recursion. Unfortunately, the recursive version runs much slower than the one presented here. The timings chart given below includes timings for an implementation of the natural recursive algorithm which was developed earlier by the authors.

The CPU timings given below were obtained on a VAX 11/780 running UNIX (Version 4.2 UCB). The notation

$$3 * \overline{1,2,\cdots,9,0}$$

means that the block 1,2,3,4,5,6,7,8,9,0, is repeated three times. The columns represent the following:

A The number of m-norm levels calculated using the recursive algorithm.

B The timing in CPU seconds for the recursive algorithm.

169

C The timing in CPU seconds for NORM.

D The level at which the m-norm stabilized.

E The norm value.

VECTOR	A	B	C	D	E
7,7,7,7,4,4,4,	4	24.45	.050	2	7.50
15,14,13, ..., 2,1	4	KILLED after one hour of connect time.	.534	3	23.75
$\overline{3*1,2,\ldots,9,0}$		6.400		3	40.50
$\overline{10*1,2,\ldots,9,0}$		651.817		3	128.75

Observe that as the length n of the vector increases, the time required by NORM to calculate the norm of the vector increases rapidly. (See column E.) In fact, it is not difficult to see that the time complexity of NORM is exponential in n. However, the time complexity of the natural recursive algorithm is exponential in both n and m, the number of levels of the m-norm calculated. On the other hand, with NORM, the time required to calculate the $(m + 1)$-norm after the m-norm has been calculated is essentially the same as the time required to calculate the $(m + 2)$-norm after the $(m + 1)$-norm has been calculated. That is, the amount of work required to calculate the m-norm of a vector using NORM is linear with respect to m.

As the recursive algorithm did not provide an easy method of determining when the m-norms stabilized, Column A gives the actual number of m-norm levels that were calculated. As a result, this algorithm provided only information about the m-norms of a vector. When the same value was obtained for the m-norm of a vector for several successive values of m, it was natural to assume that the norm of the vector equaled the repeated m-norm. However, this was only a guess, and the recursive algorithm did not seem to lend itself to a method for calculating the actual norm of a vector.

VIII. STOPPING TIME QUESTION.

Based on the problem cited in the preceding paragraph, it might seem reasonable to believe that if a vector had the same m-norm for two successive values of m, this m-norm value would be the norm value of the vector. However, it is not difficult to find vectors with finitely many nonzero terms which have an m-norm equal to an $(m + 1)$-norm, but with this m-norm value unequal to the norm value. It appears reasonable to believe that for every pair of positive integers m and k, there exists a vector $X = (x_1, x_2, \cdots, x_n)$ with $\|X\|_m = \|X\|_{m+i}$ for $1 \le i \le k$, but $\|X\|_m < \|X\|$. Therefore, if the same value is obtained for two or more successive m-norms of a vector, one cannot automatically assume this value is also the norm value of that vector. This leads to the following question:

PROBLEM 1. If $X = (x_1, x_2, \cdots, x_n)$ and k is a positive integer, find the minimal value of k (as a function of n alone) such that if m is a positive integer with $m + k \leq n$ and $\|x\|_m = \|X\|_{m+i}$ for $1 \leq i \leq k$, then $\|X\|_m = \|X\|$.

The following easy to prove fact provides a partial answer to the preceding problem.

THEOREM. If $X = (x_1, x_2, \cdots, x_n)$, then $\|X\| = \|X\|_m$ for $m \geq \lfloor (n-1)/2 \rfloor$.

Based on this result, a sufficient condition on k in the preceding problem is to take $k = \lfloor (n-1)/2 \rfloor$. However, this is possibly not a minimal value for k.

A consequence of the preceding theorem is that there exists a positive integer t such that $\|X\|_0, \|X\|_1, \cdots, \|X\|_j, \cdots$ stabilizes by the time $j = t$ for all vectors X of length n. Let $j(n)$ be the minimal value of t above.

PROBLEM 2. In the above setting,

(a) Find a reasonably tight upper bound for $j(n)$ for each positive integer n.

(b) Determine a formula for $j(n)$.

An answer to either part of Problem 2 would allow a user to estimate the time required in the worst case to evaluate the norm of a vector. Recall, the time required to calculate the $(m+1)$-norm after the m-norm has been calculated is essentially the same as the time required to calculate the $(m+2)$-norm after the $(m+1)$-norm has been calculated for all $m \geq 0$. Thus, if an upper bound for the value of $j(n)$ is k and t is the CPU time required to calculate the 1-norm of a vector after the 0-norm is calculated using the NORM program, then kt is an approximate upper bound for the CPU time needed to calculate the norm.

IX. THE TRACE PROGRAM.

The trace program allows the user to obtain some of the intermediate results used in the calculation of each m-norm for a vector. The norm program is embedded in the trace program and is available to the user when the tracing features are deactivated. The code for the trace program is roughly twice as long as the code for the norm program. Some illustrative sessions for the trace program are given below:

AN EXAMPLE OF A TRACE WITH INTERMEDIATE RESULTS PRINTED:

```
Do you want instructions? (y/n) >> n
Do you want a trace of the calculations printed? (y/n) >> y
Do you wish to have intermediate results printed? (y/n) >> y
Do you wish to have the newnorm update results printed? (y/n) >> n
```

```
Enter vector >> 12 4 6 8 10 5


 12.00  4.00  6.00  8.00 10.00  5.00
The vector has length    6
Enter maximum number of (additional) norm iterations >> 1
The current total number of norm iterations permitted is    1
Now processing vector for level =  0


VALUES OF SUP-NORMS OF SUBVECTORS
low=(row nr)    high=(column nr) >
4.00      6.00      8.00     10.00     10.00
          6.00      8.00     10.00     10.00
                    8.00     10.00     10.00
                             10.00     10.00
                                        5.00
                                       12.00
The vector norm value for this level is          0-norm =   12.00
Now processing vector for level =  1
normsums[1,.] = oldnorms[.] for parts = 1


VALUES OF NORMSUMS MATRIX FOR PARTS =   2  WITH LEVEL =   1
low=(row nr)    high=(column nr) >
0.00      0.00      0.00      0.00      0.00
          0.00     14.00     18.00     18.00
                    0.00     18.00     18.00
                              0.00     15.00
                                        0.00
Intermediate assignment B for newnorms position [  1    4] has value =   14.00
Intermediate assignment B for newnorms position [  1    5] has value =   18.00
Intermediate assignment B for newnorms position [  1    6] has value =   18.00
Intermediate assignment A for newnorms position [  2    4] has value =   14.00
Intermediate assignment A for newnorms position [  2    5] has value =   18.00
Intermediate assignment A for newnorms position [  2    6] has value =   18.00
Intermediate assignment A for newnorms position [  3    5] has value =   18.00
Intermediate assignment A for newnorms position [  3    6] has value =   18.00
```

Intermediate assignment A for newnorms position [4   6] has value =   15.00

VALUES OF NORMSUMS MATRIX FOR PARTS =   3   WITH LEVEL =   1
low=(row nr)   high=(column nr) >
0.00   0.00   0.00   0.00   0.00
          0.00   0.00   0.00   0.00
                    0.00   0.00   23.00
                              0.00   0.00
                                        0.00
Intermediate assignment B for newnorms position [1   6] has value =   23.00
Intermediate assignment B for newnorms position [2   6] has value =   23.00
Intermediate assignment A for newnorms position [3   6] has value =   23.00

UPDATE OF MATRIX NEWNORMS WITH LEVEL =   1
low=(row nr)   high=(column nr) >
4.00   6.00   8.00   10.00   11.50
          6.00   8.00   10.00   11.50
                    8.00   10.00   11.50
                             10.00   10.00
                                        5.00
                                       12.00
The vector norm value for this level is   1-norm =   12.00

Do you wish to continue processing the same vector? (y/n) >> n
Do you wish to process another vector? (y/n) >> n

Observe that in the TRACE output, values for the (lo,hi] position are printed for lo =
$1, 2, \cdots, n - 1$ and hi $= 2, 3, \cdots, n$.

AN EXAMPLE OF A TRACE WITH THE NEWNORM UPDATE RESULTS PRINTED:

Do you want instructions? (y/n) >> n
Do you want a trace of the calculations printed? (y/n) >> y
Do you wish to have intermediate results printed? (y/n) >> n
Do you wish to have the newnorm update results printed? (y/n) >> y

```
Enter vector >> 12 4 6 8 10 5


 12.00  4.00  6.00  8.00 10.00  5.00
The vector has length    6
Enter maximum number of (additional) norm iterations >> 1
The current total number of norm iterations permitted is    1
Now processing vector for level =  0


VALUES OF SUP-NORMS OF SUBVECTORS
low=(row nr)     high=(column nr) >
4.00       6.00       8.00       10.00       10.00
           6.00       8.00       10.00       10.00
                      8.00       10.00       10.00
                                 10.00       10.00
                                              5.00
                                             12.00
The vector norm value for this level is           0-norm =   12.00
Now processing vector for level =  1
normsums[1,.] = oldnorms[.] for parts = 1


VALUES OF NORMSUMS MATRIX FOR PARTS =   2  WITH LEVEL =   1
low=(row nr)     high=(column nr) >
0.00       0.00       0.00       0.00       0.00
           0.00      14.00      18.00      18.00
                      0.00      18.00      18.00
                                 0.00      15.00
                                            0.00


VALUES OF NORMSUMS MATRIX FOR PARTS =   3  WITH LEVEL =   1
low=(row nr)     high=(column nr) >
0.00       0.00       0.00       0.00       0.00
           0.00       0.00       0.00       0.00
                      0.00       0.00      23.00
                                 0.00       0.00
                                            0.00
```

174

NEWNORMS MATRIX UPDATE AT LOW = 1; HIGH = 6; PARTS = 3; NEWNORMS[1 6] =
11.50

NEWNORMS MATRIX UPDATE AT LOW = 2; HIGH = 6; PARTS = 3; NEWNORMS[2 6] =
11.50

NEWNORMS MATRIX UPDATE AT LOW = 3; HIGH = 6; PARTS = 3; NEWNORMS[3 6] =
11.50

UPDATE OF MATRIX NEWNORMS WITH LEVEL = 1
low=(row nr)   high=(column nr) >
4.00   6.00   8.00   10.00   11.50
          6.00   8.00   10.00   11.50
                    8.00   10.00   11.50
                              10.00   11.50
                              10.00   10.00
                                          5.00
                                         12.00

The vector norm value for this level is   1-norm =   12.00

Do you wish to continue processing the same vector? (y/n) >> n
Do you wish to process another vector? (y/n) >> n

   Recall that because of the physical size of a line on a terminal , TRACE produces different
styles of output for vectors of length less than 13 and those of length greater than 13. The next
two sample sessions illustrate the difference.

   AN EXAMPLE OF A SIMPLE TRACE ON A VECTOR OF LENGTH 13 OR LESS:

Do you want instructions? (y/n) >> n
Do you want a trace of the calculations printed? (y/n) >> y
Do you wish to have intermediate results printed? (y/n) >> n
Do you wish to have the newnorm update results printed? (y/n) >> n

Enter vector >> 13 12 11 10 9 8 7 6 5 4 3 2 1

```
   13.00 12.00 11.00 10.00  9.00  8.00  7.00  6.00  5.00  4.00  3.00  2.00  1.00
```
The vector has length 13
Enter maximum number of (additional) norm iterations >> 1
The current total number of norm iterations permitted is   1
Now processing vector for level = 0

VALUES OF SUP-NORMS OF SUBVECTORS
low=(row nr)   high=(column nr) >

12.00	12.00	12.00	12.00	12.00	12.00	12.00	12.00	12.00	12.00	12.00	12.00
	11.00	11.00	11.00	11.00	11.00	11.00	11.00	11.00	11.00	11.00	11.00
		10.00	10.00	10.00	10.00	10.00	10.00	10.00	10.00	10.00	10.00
			9.00	9.00	9.00	9.00	9.00	9.00	9.00	9.00	9.00
				8.00	8.00	8.00	8.00	8.00	8.00	8.00	8.00
					7.00	7.00	7.00	7.00	7.00	7.00	7.00
						6.00	6.00	6.00	6.00	6.00	6.00
							5.00	5.00	5.00	5.00	5.00
								4.00	4.00	4.00	4.00
									3.00	3.00	3.00
										2.00	2.00
											1.00
											13.00

The vector norm value for this level is   0-norm =   13.00
Now processing vector for level = 1
normsums[1,.] = oldnorms[.] for parts = 1

VALUES OF NORMSUMS MATRIX FOR PARTS =   2 WITH LEVEL =   1
low=(row nr)   high=(column nr) >

0.00	0.00	0.00	0.00	0.00	0.00	0.00	0.00	0.00	0.00	0.00	0.00
	0.00	21.00	21.00	21.00	21.00	21.00	21.00	21.00	21.00	21.00	21.00
		0.00	19.00	19.00	19.00	19.00	19.00	19.00	19.00	19.00	19.00
			0.00	17.00	17.00	17.00	17.00	17.00	17.00	17.00	17.00
				0.00	15.00	15.00	15.00	15.00	15.00	15.00	15.00
					0.00	13.00	13.00	13.00	13.00	13.00	13.00
						0.00	11.00	11.00	11.00	11.00	11.00
							0.00	9.00	9.00	9.00	9.00
								0.00	7.00	7.00	7.00
									0.00	5.00	5.00
										0.00	3.00
											0.00

VALUES OF NORMSUMS MATRIX FOR PARTS =   3 WITH LEVEL =   1

low=(row nr)   high=(column nr) >

```
   0.00   0.00   0.00   0.00   0.00   0.00   0.00   0.00   0.00   0.00   0.00   0.00
          0.00   0.00   0.00   0.00   0.00   0.00   0.00   0.00   0.00   0.00   0.00
                 0.00   0.00  27.00  27.00  27.00  27.00  27.00  27.00  27.00  27.00
                        0.00   0.00  24.00  24.00  24.00  24.00  24.00  24.00  24.00
                               0.00   0.00  21.00  21.00  21.00  21.00  21.00  21.00
                                      0.00   0.00  18.00  18.00  18.00  18.00  18.00
                                             0.00   0.00  15.00  15.00  15.00  15.00
                                                    0.00   0.00  12.00  12.00  12.00
                                                           0.00   0.00   9.00   9.00
                                                                  0.00   0.00   6.00
                                                                         0.00   0.00
                                                                                0.00
```

CURRENT NORM OF VECTOR updated to   13.50

VALUES OF NORMSUMS MATRIX FOR PARTS =   4   WITH LEVEL =   1

low=(row nr)   high=(column nr) >

```
   0.00   0.00   0.00   0.00   0.00   0.00   0.00   0.00   0.00   0.00   0.00   0.00
          0.00   0.00   0.00   0.00   0.00   0.00   0.00   0.00   0.00   0.00   0.00
                 0.00   0.00   0.00   0.00   0.00   0.00   0.00   0.00   0.00   0.00
                        0.00   0.00   0.00  30.00  30.00  30.00  30.00  30.00  30.00
                               0.00   0.00   0.00  26.00  26.00  26.00  26.00  26.00
                                      0.00   0.00   0.00  22.00  22.00  22.00  22.00
                                             0.00   0.00   0.00  18.00  18.00  18.00
                                                    0.00   0.00   0.00  14.00  14.00
                                                           0.00   0.00   0.00  10.00
                                                                  0.00   0.00   0.00
                                                                         0.00   0.00
                                                                                0.00
```

CURRENT NORM OF VECTOR updated to   15.00

VALUES OF NORMSUMS MATRIX FOR PARTS =   5   WITH LEVEL =   1

low=(row nr)   high=(column nr) >

```
0.00   0.00   0.00   0.00   0.00   0.00   0.00   0.00   0.00   0.00   0.00   0.00
       0.00   0.00   0.00   0.00   0.00   0.00   0.00   0.00   0.00   0.00   0.00
              0.00   0.00   0.00   0.00   0.00   0.00   0.00   0.00   0.00   0.00
                     0.00   0.00   0.00   0.00   0.00   0.00   0.00   0.00   0.00
                            0.00   0.00   0.00   0.00  30.00  30.00  30.00  30.00
                                   0.00   0.00   0.00   0.00  25.00  25.00  25.00
                                          0.00   0.00   0.00   0.00  20.00  20.00
                                                 0.00   0.00   0.00   0.00  15.00
                                                        0.00   0.00   0.00   0.00
                                                               0.00   0.00   0.00
                                                                      0.00   0.00
                                                                             0.00
```

VALUES OF NORMSUMS MATRIX FOR PARTS =   6   WITH LEVEL =   1

low=(row nr)   high=(column nr) >

```
0.00   0.00   0.00   0.00   0.00   0.00   0.00   0.00   0.00   0.00   0.00   0.00
       0.00   0.00   0.00   0.00   0.00   0.00   0.00   0.00   0.00   0.00   0.00
              0.00   0.00   0.00   0.00   0.00   0.00   0.00   0.00   0.00   0.00
                     0.00   0.00   0.00   0.00   0.00   0.00   0.00   0.00   0.00
                            0.00   0.00   0.00   0.00   0.00   0.00   0.00   0.00
                                   0.00   0.00   0.00   0.00   0.00  27.00  27.00
                                          0.00   0.00   0.00   0.00   0.00  21.00
                                                 0.00   0.00   0.00   0.00   0.00
                                                        0.00   0.00   0.00   0.00
                                                               0.00   0.00   0.00
                                                                      0.00   0.00
                                                                             0.00
```

UPDATE OF MATRIX NEWNORMS WITH LEVEL =   1

low=(row nr)   high=(column nr) >

```
12.00  12.00  12.00  12.00  13.50  13.50  15.00  15.00  15.00  15.00  15.00  15.00
       11.00  11.00  11.00  13.50  13.50  15.00  15.00  15.00  15.00  15.00  15.00
              10.00  10.00  13.50  13.50  15.00  15.00  15.00  15.00  15.00  15.00
                      9.00   9.00  12.00  15.00  15.00  15.00  15.00  15.00  15.00
                             8.00   8.00  10.50  13.00  15.00  15.00  15.00  15.00
                                    7.00   7.00   9.00  11.00  12.50  13.50  13.50
                                           6.00   6.00   7.50   9.00  10.00  10.50
                                                  5.00   5.00   6.00   7.00   7.50
                                                         4.00   4.00   4.50   5.00
                                                                3.00   3.00   3.00
                                                                       2.00   2.00
                                                                              1.00
                                                                             15.00
```

The vector norm value for this level is   1-norm =   15.00

178

Do you wish to continue processing the same vector? (y/n) >> n
Do you wish to process another vector? (y/n) >> n

The next example was aborted with a C after the first matrix output. All other output would be changed in the same manner. Since the same vector was used with only an additional zero at the end to force the length to be 14, the output format for this example can be compared easily to the previous one.

AN EXAMPLE OF THE FIRST MATRIX OUTPUT OF A SIMPLE
TRACE FOR A VECTOR OF LENGTH 14 OR GREATER:

Do you want instructions? (y/n) >> n
Do you want a trace of the calculations printed? (y/n) >> y
Do you wish to have intermediate results printed? (y/n) >> n
Do you wish to have the newnorm update results printed? (y/n) >> n

Enter vector >> 13 12 11 10 9 8 7 6 5 4 3 2 1 0

 13.00 12.00 11.00 10.00 9.00 8.00 7.00 6.00 5.00 4.00 3.00 2.00 1.00
0.00
The vector has length 14
Enter maximum number of (additional) norm iterations >> 2
The current total number of norm iterations permitted is   2
Now processing vector for level = 0

VALUES OF SUP-NORMS OF SUBVECTORS

[1 2]=	12.00	[1 3]=	12.00	[1 4]=	12.00	[1 5]=	12.00
[1 6]=	12.00	[1 7]=	12.00	[1 8]=	12.00	[1 9]=	12.00
[1 10]=	12.00	[1 11]=	12.00·	[1 12]=	12.00	[1 13]=	12.00
[1 14]=	12.00	[2 3]=	11.00	[2 4]=	11.00	[2 5]=	11.00
[2 6]=	11.00	[2 7]=	11.00	[2 8]=	11.00	[2 9]=	11.00
[2 10]=	11.00	[2 11]=	11.00	[2 12]=	11.00	[2 13]=	11.00
[2 14]=	11.00	[3 4]=	10.00	[3 5]=	10.00	[3 6]=	10.00
[3 7]=	10.00	[3 8]=	10.00	[3 9]=	10.00	[3 10]=	10.00
[3 11]=	10.00	[3 12]=	10.00	[3 13]=	10.00	[3 14]=	10.00
[4 5]=	9.00	[4 6]=	9.00	[4 7]=	9.00	[4 8]=	9.00
[4 9]=	9.00	[4 10]=	9.00	[4 11]=	9.00	[4 12]=	9.00
[4 13]=	9.00	[4 14]=	9.00	[5 6]=	8.00	[5 7]=	8.00
[5 8]=	8.00	[5 9]=	8.00	[5 10]=	8.00	[5 11]=	8.00
[5 12]=	8.00	[5 13]=	8.00	[5 14]=	8.00	[6 7]=	7.00
[6 8]=	7.00	[6 9]=	7.00	[6 10]=	7.00	[6 11]=	7.00
[6 12]=	7.00	[6 13]=	7.00	[6 14]=	7.00	[7 8]=	6.00
[7 9]=	6.00	[7 10]=	6.00	[7 11]=	6.00	[7 12]=	6.00
[7 13]=	6.00	[7 14]=	6.00	[8 9]=	5.00	[8 10]=	5.00
[8 11]=	5.00	[8 12]=	5.00	[8 13]=	5.00	[8 14]=	5.00
[9 10]=	4.00	[9 11]=	4.00	[9 12]=	4.00	[9 13]=	4.00
[9 14]=	4.00	[10 11]=	3.00	[10 12]=	3.00	[10 13]=	3.00
[10 14]=	3.00	[11 12]=	2.00	[11 13]=	2.00	[11 14]=	2.00
[12 13]=	1.00	[12 14]=	1.00	[13 14]=	0.00	[14 14]=	13.00

The vector norm value for this level is   0-norm =   13.00

^C

To activate the trace, the user must reply y (for yes) when the question "Do you want a trace of the calculations printed? (y/n) >>" appears. As indicated in the sample session, the user will next be asked whether or not intermediate results should be printed and whether or not the NEWNORMS update results should be printed. The effect of answer y (for yes) for each of these choices is illustrated by the examples above and will be discussed later in this section. However, both options are included primarily for the user who wants to follow the details of the calculations very closely.

The first time it is invoked, the procedure NextLevel will print the message "VALUE OF SUP-NORM OF SUBVECTORS" and will call procedure Display1 to print the values stored in NEWNORMS. If the value of SMALL is true, the results will be displayed in the matrix form discussed in Section III. Otherwise, the results will be printed in four columns in the form

[LO,HI] = newnorms-value

The value in NEWNORMS corresponds to the sup-norm value of the subvector (LO,HI] which was discussed in Section III.

Currently the setting of SMALL is TRUE if the vector entered by the user has length less than 14. The choice of 14 was made because the procedures Display1 and Display2 can fit their matrix form of output for vectors of length less than 14 into the 80 columns normally provided on a terminal or monitor screen. However, this form also assumes that the integer part of any number output by Display1 and Display2 has only two digits. To allow for larger integral values, the size of the output fields used by these two procedures must be enlarged. If the size of the output fields are enlarged, then fewer than 13 fields can fit into 80 columns and it will be necessary to change the definition of SMALL appropriately.

On subsequent calls, procedure NextLevel calls procedure SumsOfNorms. If no entry in NORMSUMS increases during the calculation of NORMSUMS[(PARTS+1) mod 2, .] from NORMSUMS[PARTS mod 2, .], SUMSGROW is FALSE and the message

NORMSUMS MATRICES STABILIZED AT THIS

LEVEL WITH PARTS = _____

is printed. Otherwise, the message

VALUES OF NORMSUMS MATRIX FOR PARTS = _____ WITH LEVEL = _____

is printed and procedure Display2 is called to print NORMSUMS[(PARTS+1) mod 2, .]. The output format and interpretation of this output is the same as for procedure Display1, discussed above.

If SUMSGROW is TRUE, then procedure UpdateNorm is called by procedure SumsOfNorms. If the user elected to have intermediate results printed, then on each of the calculations described in Steps (20) - (21) of the algorithm in Section IV, the following message is printed:

INTERMEDIATE ASSIGNMENT_____ FOR NORM POSITION [   ]

HAS VALUE_____.

This option will normally generate a large amount of output and should not be used unless these calculations are needed. Redirecting the output to a file would be useful. This can be accomplished by anyone who knows the file management system for the computer on which you are running the program.

Continuing with procedure UpdateNorm, if the user decided to have updates to NEWNORMS reported, then each time a value in NEWNORMS is replaced with a larger value (as described in step 22 of the algorithm in Section IV), the message

NEWNORMS MATRIX UPDATE AT LOW =_____, HIGH =_____,

PARTS =_____; [LO,HI] =_____

is printed. This option will not generate as much output as the "intermediate results" option.

Independent of the user's choice on the "intermediate results" and the "updates to NEWNORMS options, procedure UpdateNorm will print certain information for tracing. After NEWNORMS is recalculated, if a change has occurred that increases the norm of the current vector, then the message

CURRENT NORM OF VECTOR UPDATED TO _____

is printed.

When procedures UpdateNorm and SumsOfNorms terminate, procedure NextLevel is reactivated. It will print the message

UPDATE OF MATRIX NEWNORMS WITH LEVEL = _____

It then calls procedure Display1 to print the current values of NEWNORMS using the format described earlier.

Finally, control returns to procedure ProcessVector and the message

THE VECTOR NORM FOR THIS LEVEL IS

_____-LEVEL =_____

If the calculations are not complete, procedure ProcessVector will start the entire process again.

X. PASCAL CODE FOR THE NORM PROGRAM.

```
program norm (input,output);
```

(* This program calculates the Tsirelson's space norm for vectors of finite length.

The algorithm was developed by Johnnie W. Baker (Kent State University, Kent, Ohio) and Oberta A. Slotterbeck (Hiram College, Hiram, Ohio) with mathematical support from Richard Aron (Kent State University, Kent, Ohio, and Trinity College, Dublin, Ireland). The original suggestion of developing a computer algorithm to calculate the Tsirelson's space norm was due to Richard Aron. Tim Murphy (Trinity College, Dublin, Ireland) provided some speed-up features that appear in this version.

The bound on the length of the vector is MAXDIM. The vectors used for storage in this program have length SIZESTORE which must be at least

MAXDIM * (MAXDIM - 1)/2 + 1

The values of MAXDIM and SIZESTORE must be increased to process vectors longer than the current value of MAXDIM. Their values can be decreased if the amount of storage required for execution is too large for your computer.

Field widths for output are small and can be increased if necessary. *)

```
const
    maxdim  = 100;
    sizestore = 4960;
var
    vector:    array[1..maxdim] of real;
    level, toplevel, dim, last :  integer;
    answer:                    char;
procedure instructions;
    begin
    writeln;
    writeln('Enter vector components, separated by one or more spaces' );
    writeln('following the \*'Enter vector >> \*' prompt. For example:');
    writeln('Enter vector >> 4 4 4 7 7 7 7');
    writeln('If you can not enter all of the components on one line,');
    writeln('continue entering the components on the next line, either by');
    writeln('running off the end and continuing, or by hitting the return');
    writeln('key immediately after a space has been entered.');
    writeln('DO NOT enter a space after the last vector component.');
    writeln('When you are instructed to enter (y/n), enter \*'y\*' for yes');
    writeln('and \*'n\*' for no. DO NOT enter a space before the \*'y\*' or \*'n\*'
    writeln;
    end;
procedure EnterVector;

(* This procedure permits the user to enter a vector of length not greater than "maxdim" ". If
    vector is too long, an error message is printed and the user is prompted to re-enter another
    vector. *)

    begin
    dim := 0;
    writeln;
    write('Enter vector >> ');
    readln;
    while not eoln do
        begin
        dim := dim+1;
        read(vector[dim]);
```

```
    end;
  writeln;
  if dim > maxdim then
    begin
    writeln('ERROR! Length of vector exceeds ',maxdim:4 );
    writeln('value of MAXDIM must be increased to ', dim:4 ,
                ' to process this vector');
    end;
  end;  (* of EnterVector *)
```

procedure ProcessVector;

(* This procedure calculates the m-norms of current vector entered by the user. It reports the value of each m-norm as they are calculated, for $m = 0$ up to the maximum number of levels currently allowed by the user. If the m-norm values stabilize, this fact is reported and the procedure terminates (i.e., returns). *)

```
var
    oldnorms,newnorms:  array[1..sizestore] of real;
    normsums:           array[0..1,1..sizestore] of real;
    oldlevel:                integer;
    stabilized:              boolean;
```

procedure NextLevel;

(* This procedure calculates the next level of m-norms for the subvectors for the vector entered by the user (i.e., in this setting, level-norm is calculated.) *)

```
var
    lo, mid, hi, index:  integer;
    x:                   real;
```

procedure SumsOfNorms;

(* This procedure calculates values for norm sums by repeatedly computing the new values for (PARTS+1) in NORMSUM[(PARTS+1)mod2 , .] from NORMSUMS[(PARTS)mod 2 , .] . The values of newnorms is developed also by repeated calls to UpdateNorm *)

```
var
```

184

```
   index, indexA, indexB: integer;
   hi, lo, mid:                integer;
   parts, partplus:            integer;
   y:                          real;
   sumsgrow:                   boolean;


procedure UpdateNorm;

(* Updates newnorms[.] for (parts+1) using the new values calculated in procedure Sum-
   sOfNorms for normsums[(parts+1)mod 2 , . ]. *)

var
   hi, lo : integer;

begin
for lo := 1 to dim-1 do
   for hi := lo+1 to dim do
      begin
      index := (lo-1)*dim + hi - (lo*(lo+1))div 2;
      x := 0;
      if (partplus <= hi-lo) and (partplus <= lo) then (* use this assignment
               for value of x *)
         x := normsums[(partplus)mod 2 , index];


      if (partplus > lo) and (partplus <= hi div 2) then (*use this assignment
      for value of x *)
         begin
         indexA := parts*dim + hi - (partplus*(parts+2))div 2;
         x := normsums[(partplus)mod 2 , indexA];
         end;
      x := x/2;
      if newnorms[index] < x then  (* update newnorms vector using value x *)
         begin
         stabilized := false;
         newnorms[index] := x;
         end
```

185

```
        end;
if newnorms[last] < newnorms[dim-1] then     (* update newnorms value  *)
    newnorms[last] := newnorms[dim-1];
end; (* UpdateNorm *)


begin (* SumOfNorms *)
sumsgrow := true;
for parts := 1 to  (dim div 2)-1  do
    if (sumsgrow) then
        begin
        partplus := parts + 1;
        sumsgrow := false;
        for lo := partplus to dim-1 do
            for hi := partplus+lo to dim do
                if parts < hi-lo then
                    begin
                    x := 0;
                    for mid := lo+1 to hi-parts do
                        begin
                        indexA := (lo-1)*dim + mid - (lo*(lo+1))div 2;
                        indexB := (mid-1)*dim + hi - (mid*(mid+1))div 2;
                        y := oldnorms[indexA] + normsums[(parts)mod 2 , indexB];
                        if x < y then x := y;
                        end;
                    index := (lo-1)*dim + hi - (lo*(lo+1))div 2;
                    normsums[(partplus)mod 2 , index] := x;

                    y := normsums[(parts)mod 2 , index]; (*Has a larger value been*)
                    if x > y then sumsgrow := true;      (* found for NORMSUMS    *)
                    end;                                 (* for PARTS+1?          *)


        if sumsgrow then
            UpdateNorm;
        end;
end; (* SumsOfNorms *)
```

```
begin    (*  NextLevel  *)


if level = 0 then    (* store sup-norm of subvectors in the norm vector *)
   begin
   for lo := 1 to dim-1 do
       for hi := lo+1 to dim do
          begin
          x := 0;
          for mid := lo+1 to hi do
             if abs(vector[mid]) > x then x := abs(vector[mid]);
          index := (lo-1)*dim + hi - (lo*(lo+1))div 2;
          newnorms[index] := x;
          end;
   stabilized := false;
   if newnorms[dim-1] > abs(vector[1]) then
       newnorms[last] := newnorms[dim-1]
   else newnorms[last] := abs(vector[1]);   (* 0-norm stored in newnorms[last]*)
   for index := 1 to last do                     (* Copy NEWNORMS into OLDNORMS    *)
       oldnorms[index] := newnorms[index];
   end
else    (*    level > 0    *)
   begin

   (* First, for parts = 1, set normsums[1,.] =  oldnorms[.]  *)

   for index := 1 to last do
       normsums[1,index] := oldnorms[index];

   (* Next call SumsOfNorms to repeatedly compute the new values for

(parts+1) in normsums[(parts+1)mod 2,.] from normsums[(parts)mod 2,.]  *)

   SumsOfNorms;

   end;
   for index := 1 to last do                     (* Copy NEWNORMS into OLDNORMS    *)
```

```
      oldnorms[index] := newnorms[index];
end;   (*  NextLevel  *)
begin (*  ProcessVector  *)


oldlevel := toplevel;
write('Enter maximum number of (additional) norm iterations >> ' );
if eoln then readln;
read(toplevel);
toplevel := toplevel + oldlevel;
stabilized := false;
while (level <= toplevel) and (not stabilized) do
   begin
   stabilized := true;
   NextLevel;
   writeln( level, '-norm = ', newnorms[last]:7:2);
   writeln;
   if stabilized then
      writeln( 'The norms stabilized after ', level:3 , ' iterations');
   level := level + 1;
   end;
if stabilized then answer := 'n' else answer := 'y';
if not stabilized then
   begin
   write('Do you wish to continue processing the same vector? (y/n) >> ');
   if eoln then readln;
   read(answer);
   if answer <> 'n' then answer := 'y';
   end
end;   (*  ProcessVector  *)


begin  (*  main line  *)


answer := 'y';
write('Do you want instructions? (y/n) >> ' );
if eoln then readln;
read(answer);
```

```
if answer <> 'n' then
    instructions;

repeat
    EnterVector;
    toplevel := 0;
    level := 0;
    last := (dim*(dim-1))div 2 +1;
    answer := 'y';
    while ((answer='y') and (dim<>0) and (dim<=maxdim)) do
        ProcessVector;
    answer := 'y';
    write('Do you wish to process another vector? (y/n) >> ');
    if eoln then readln;
    read(answer);
until (answer = 'n');
end.  (*  of main line  *)
```

X1. PASCAL CODE FOR THE TRACE PROGRAM.

```
program trace (input,output);
```

(* This program calculates the Tsirelson's space norm for vectors of finite length and provides tracing information for the user.

The algorithm was developed by Johnnie W. Baker (Kent State University, Kent, Ohio) and Oberta A. Slotterbeck (Hiram College, Hiram, Ohio) with mathematical support from Richard Aron (Kent State University, Kent, Ohio, and Trinity College, Dublin, Ireland). The original suggestion of developing a computer algorithm to calculate the Tsirelson's space norm was due to Richard Aron. Tim Murphy (Trinity College, Dublin, Ireland) provided some speed-up features that appear in this version.

The bound on the length of the vector is MAXDIM. The vectors used for storage in this program have length SIZESTORE which must be at least

$$MAXDIM * (MAXDIM - 1)/2 + 1$$

The values of MAXDIM and SIZESTORE must be increased to process vectors longer than the current value of MAXDIM. Their values can be decreased if the amount of storage required for execution is too large for your computer.

Field widths for output are small and can be increased if necessary. *)

```
const
```

```
    maxdim   = 100;
    sizestore = 4960;
  var
    vector:   array[1..maxdim] of real;
    level, toplevel, dim, last :  integer;
    trace, small :                boolean;
    answer, intermediate, update :char;


  procedure instructions;
    begin
    writeln;
    writeln('Enter vector components, separated by one or more spaces' );
    writeln('following the \*'Enter vector >> \*' prompt. For example:');
    writeln('Enter vector >> 4 4 4 7 7 7 7');
    writeln('If you can not enter all of the components on one line,');
    writeln('continue entering the components on the next line, either by');
    writeln('running off the end and continuing, or by hitting the return');
    writeln('key immediately after a space has been entered.');
    writeln('DO NOT enter a space after the last vector component.');
    writeln('When you are instructed to enter (y/n), enter \*'y\*' for yes');
    writeln('and \*'n\*' for no. DO NOT enter a space before the \*'y\*' or \*'n\*'
    writeln;
    end;


procedure EnterVector;

(* This procedure permits the user to enter a vector of length not greater than "maxdim". If
   vector is too long, an error message is printed and the user is prompted to re-enter another
   vector. *)

var
    index: integer;

    begin
    dim := 0;
    writeln;
    write('Enter vector >> ');
```

190

```
      readln;
      while not eoln do
          begin
          dim := dim+1;
          read(vector[dim]);
          end;
      writeln;
      if dim > maxdim then
          begin
          writeln('ERROR! Length of vector exceeds ',maxdim:4 );
          writeln('value of MAXDIM must be increased to ', dim:4 ,
                        ' to process this vector');
          end;
      if trace and (dim <= maxdim) then
          begin
          for index := 1 to dim do
              write(vector[index]:6:2);
          writeln;
          writeln('The vector has length ',dim:3 );
          end;
      end;   (* of EnterVector *)


procedure ProcessVector;

(* This procedure calculates the m-norms of current vector entered by the user. It reports the
   value of each $m$-norm as they are calculated, for $m = 0$ up to the maximum number of
   levels currently allowed by the user. If the $m$-norm values stabilize, this fact is reported
   and the procedure terminates (i.e., returns). *)

var
    oldnorms, newnorms:     array[1..sizestore] of real;
    normsums:               array[0..1,1..sizestore] of real;
    oldlevel:                 integer;
    stabilized:             boolean;


procedure NextLevel;

(* This procedure calculates the next level of $m$-norms for the subvectors for the vector entered
```

191

by the user (i.e., in this setting level-norm is calculated. *)

```
var
  lo, mid, hi:            integer;
  index, indexA, column : integer;
  x:                      real;
```

procedure Display1;

(* This procedure is used only if the trace switch is on. It prints the values stored in the vector NEWNORMS in matrix form if the current length of the vector stored in VECTQR (i.e.,the value of DIM) does not exceed SMALL. SMALL has been preassigned the value 13. Otherwise, the values in NORM are printed in four columns. *)

```
var
   hi, lo: integer;
begin
if small then
   begin
       writeln( 'low=(row nr)      high=(column nr) >');
       for lo := 1 to (dim -1) do
          begin
          writeln;
          for hi := 2 to dim do
             if lo < hi then
                begin
                index := (lo - 1)*dim + hi - (lo * (lo + 1))div 2;
                write(newnorms[index]:6:2)
                end
             else write('      ');
          end;
   writeln;
   for hi := 2 to (dim-1) do
       write (' ':6);
   index := dim * (dim - 1) div 2 + 1;
   write ( newnorms[index]:6:2);
   writeln;
       writeln;
```

```
        end
else
    begin
    column := 0;
    for lo := 1 to dim do
        for hi := (lo + 1) to dim do
            begin
            index := (lo-1)*dim + hi - (lo*(lo+1))div 2;
            write('[', lo:3, ' ', hi:3, ']= ', newnorms[index]:7:2, '   ');
            column := column + 1;
            column := (column) mod 4;
            if column = 0 then writeln;
            end;
  index := dim * (dim - 1) div 2 + 1;
  write ('[', dim:3, ' ', dim:3, '] =' , newnorms[index]:7:2);
  writeln;
    writeln;
    end;
end;        (* of Display1  *)


procedure Display2;

(* This procedure is used only if the trace switch is on. It prints the values stored in the vector
    NORMSUMS corresponding to the current value of PARTS. NORMSUMS is being used
    to store a three dimensional matrix. These values are printed in matrix form if the current
    length of the vector stored in VECTOR (i.e.,the value of DIM) does not exceed SMALL.
    SMALL has been preassigned the value 13. Otherwise, the values in NORMSUMS for
    current value of PARTS are printed in four columns. Code is similar to that in Display1.
    *)

var
    hi, lo : integer;
begin
if small then
    begin
    writeln( 'low=(row nr)     high=(column nr) >');
    for lo := 1 to (dim -1) do
```

193

```
      begin
      writeln;
      for hi := 2 to dim do
         if lo < hi then
            begin
            index := (lo - 1)*dim + hi - (lo * (lo + 1))div 2;
            write(normsums[indexA,index]:6:2)
            end
         else write('        ');
      end;
   writeln;
   writeln;
   end
else
   begin
   column := 0;
   for lo := 1 to dim do
      for hi := (lo + 1) to dim do
         begin
         index := (lo-1)*dim + hi - (lo*(lo+1))div 2;
         write('[', lo:3, ' ',hi:3, ']= ',normsums[indexA,index]:7:2, '   ');
         column := column + 1;
         column := (column)mod 4;
         if column = 0 then writeln;
         end;
   writeln;
   writeln;
   end;
end; (* Display2 *)


procedure SumsOfNorms;

(* This procedure calculates values for NORMSUMS by repeatedly computing the new values
    for (PARTS+1) in NORMSUMS[ (PARTS+1)mod2 , . ] from NORMSUMS[ (PARTS)mod
    2 , . ]. The values of NEWNORMS is developed also by repeated calls to UpdateNorm. *)

var
```

```
   index, indexB:      integer;
   hi, lo, mid:          integer;
   parts, partplus:   integer;
   y:                          real;
   sumsgrow:             boolean;


procedure UpdateNorm;

(* Updates newnorms[.]  for (parts+1) using the new values calculated in procedure Sum
   sOfNorms for normsums[(parts+1)mod 2, . ]. *)

var
   hi, lo : integer;
begin
for lo := 1 to dim-1 do
   for hi := lo+1 to dim do
       begin
       index := (lo-1)*dim + hi - (lo*(lo+1))div 2;
       x := 0;
       if (partplus <= hi-lo) and (partplus <= lo) then (* Use assignment A
                                                     for value of x  *)
          begin
          x := normsums[(partplus)mod 2 , index];
          if trace then
             if intermediate = 'y' then
                writeln('Intermediate assignment A for newnorms position ',
                       '[',lo:3,' ',hi:3, '] has value = ', x:7:2);
          end;
       if (partplus > lo) and (partplus <= hi div 2) then   (* Use assignment B
                                                     for value of x *)
          begin
          indexA := parts*dim + hi - (partplus*(parts+2))div 2;
          x := normsums[(partplus)mod 2 , indexA];
          if trace then
             if intermediate = 'y' then
                writeln('Intermediate assignment B for newnorms position ',
'[',lo:3,' ',hi:3, '] has value = ', x:7:2);
```

```
            end;
        x := x/2;
        if newnorms[index] < x then   (* update norm vector using value x *)
            begin
            stabilized := false;
            newnorms[index] := x;
            if trace then
                if update = 'y' then
                    begin
                    writeln('NEWNORMS MATRIX UPDATE AT LOW = ',lo:3,'; HIGH = '
                            hi:3);
            writeln('PARTS = ':32,partplus:3);
                    writeln('; NEWNORMS[',lo:3,' ',hi:3,'] = ',x:7:2);
                    end
            end
        end;
if newnorms[last] < newnorms[dim-1] then    (* update newnorms value *)
    begin
    newnorms[last] := newnorms[dim-1];
    if trace then
        begin
        writeln;
        writeln('CURRENT NORM OF VECTOR updated to ', newnorms[last]:7:2);
        writeln
        end
    end;
end; (* UpdateNorm *)


begin (* SumsOfNorms *)
sumsgrow := true;
for parts := 1 to  (dim div 2)-1  do
    if (sumsgrow) then
        begin
        partplus := parts + 1;
        sumsgrow := false;
        if trace then
```

```
      for index := 1 to last do  (* zero the matrix before storage *)
          normsums[(partplus)mod 2,index] := 0.0;
  for lo := partplus to dim-1 do
      for hi := partplus+lo to dim do
          if parts < hi-lo then
              begin
              x := 0;
              for mid := lo+1 to hi-parts do
                  begin
                  indexA := (lo-1)*dim + mid - (lo*(lo+1))div 2;
                  indexB := (mid-1)*dim + hi - (mid*(mid+1))div 2;
                  y := oldnorms[indexA] + normsums[(parts)mod 2 , indexB];
                  if x < y then x := y;
                  end;
              index := (lo-1)*dim + hi - (lo*(lo+1))div 2;
              normsums[(partplus)mod 2 , index] := x;

              y := normsums[(parts)mod 2 , index]; (*Has a larger value been *)
              if (x > y) then sumsgrow := true;    (* found for NORMSUMS      *)
              end;                                 (* for PARTS+1?            *)

  if trace and (not sumsgrow) then        (* no larger values computed   *)
                                          (* for NORMSUMS for PARTS+1    *)
      writeln('NORMSUMS matrices stabilized for this level',
          ' with parts = ', parts:3);

  if trace and sumsgrow then
      begin
      writeln;
      writeln('VALUES OF NORMSUMS MATRIX FOR PARTS = ', partplus:3,
                              ' WITH LEVEL = ', level:3);
      indexA := (partplus)mod 2;
      Display2;
      end;
  if sumsgrow then
      UpdateNorm;
```

```
      end;
end; (*  SumsOfNorm  *)


begin   (*  NextLevel  *)


if level = 0 then    (* store sup-norm of subvectors in the norm vector *)
   begin
   for lo := 1 to dim-1 do
      for hi := lo+1 to dim do
         begin
         x := 0;
         for mid := lo+1 to hi do
            if abs(vector[mid]) > x then x := abs(vector[mid]);
         index := (lo-1)*dim + hi - (lo*(lo+1))div 2;
         newnorms[index] := x;
         end;
   stabilized := false;
   if newnorms[dim-1] > abs(vector[1]) then
      newnorms[last] := newnorms[dim-1]
   else newnorms[last] := abs(vector[1]); (*0-norm of vector stored in *)
                                    (*  newnorms[last]            *)
   for index := 1 to last do               (* Copy NEWNORMS into OLDNORMS *)
      oldnorms[index] := newnorms[index];
   if trace then
      begin
      writeln;
      writeln('VALUES OF SUP-NORMS OF SUBVECTORS');
      Display1;
      writeln
      end;
   end
else     (*   level > 0   *)
   begin

   (* First, for parts = 1, set normsums[1,.] = oldnorms[.]  *)
```

198

```pascal
      for index := 1 to last do
         normsums[1,index] := oldnorms[index];
      if trace then
         begin
         writeln('normsums[1,.] = oldnorms[.] for parts = 1');
         writeln;
         end;


   (* Next call SumsOfNorms to repeatedly compute the new values for

      (parts+1) in normsums[(parts+1)mod 2,.] from normsums[(parts)mod 2,.]  *)


   SumsOfNorms;

      if trace then
         begin
         writeln;
         writeln('UPDATE OF MATRIX NEWNORMS WITH LEVEL = ', level:3);
         Display1;
         end
      end;
      for index := 1 to last do            (* Copy NEWNORMS into OLEDNORMS *)
         oldnorms[index] := newnorms[index];
end;    (* NextLevel *)


begin  (* ProcessVector *)

oldlevel := toplevel;
write('Enter maximum number of (additional) norm iterations >> ' );
if eoln then readln;
read(toplevel);
toplevel := toplevel + oldlevel;
if trace then
   writeln('The current total number of norm iterations',
                     ' permitted is ',  toplevel:3 );
stabilized := false;
```

```
while (level <= toplevel) and (not stabilized) do
   begin
   stabilized := true;
   if trace then
      begin
      writeln('Now processing vector for level = ', level:2);
      writeln;
      end;
   NextLevel;
   if trace then
      write('The vector norm value for this level is ');
   writeln( level, '-norm = ', newnorms[last]:7:2);
   writeln;
   if stabilized then
      writeln( 'The norms stabilized after ', level:3 , ' iterations');
   level := level + 1;
   end;
if stabilized then answer := 'n' else answer := 'y';
if not stabilized then
   begin
   write('Do you wish to continue processing the same vector? (y/n) >> ');
   if eoln then readln;
   read(answer);
   if answer <> 'n' then answer := 'y';
   end
end;   (*  ProcessVector  *)


begin  (*  main line  *)

answer := 'y';
write('Do you want instructions? (y/n) >> ' );
if eoln then readln;
read(answer);
if answer <> 'n' then
   instructions;
repeat
```

```
    answer := 'n';
    write('Do you want a trace of the calculations printed? (y/n) >> ');
    if eoln then readln;
    read(answer);
    if answer <> 'y' then trace := false else trace := true;
    intermediate := 'n';
    if trace then
        begin
        write('Do you wish to have intermediate results printed? (y/n) >> ');
        if eoln then readln;
        read(intermediate);
        update := 'n';
        write('Do you wish to have the newnorm update results printed? (y/n) >> ');
        if eoln then readln;
        read(update);
        end;
    EnterVector;
    toplevel := 0;
    level := 0;
    last := (dim*(dim-1))div 2 +1;
    if dim > 13 then small := false else small := true;
    answer := 'y';
    while ((answer='y') and (dim<>0) and (dim<=maxdim)) do
        ProcessVector;
    answer := 'y';
    write('Do you wish to process another vector? (y/n) >> ');
    if eoln then readln;
    read(answer);
    until (answer = 'n');
end. (* of main line *)
```

Appendix: Notes and Remarks.

1. The programs above which calculate the norm in Tsirelson's's space were designed around the Figiel-Johnson definition of admissible sums: $k < E_1 < \cdots < E_k$. Our definition in Construction I.1(e) allows $k \leq E_1 < \cdots < E_k$. It's immediately clear that $\{t_n\}_{n=1}^{\infty}$ under our definition is isometrically equivalent to $\{t_n\}_{n=2}^{\infty}$ under the definition of [26] and [35]. So to calculate $\|(2,0,2,0,1,0,0,\cdots)\|$ under our definition, simply input $(0,2,0,2,0,1,0,0,\cdots)$

into the program, etc.

2. The extreme points of B_T:

These programs were used to find out some facts about the nature of extreme points in T, all of which were later proven. A few computations show that although the basis $\{t_n\}_{n=1}^{\infty}$ is 1-unconditional, it fails to be strictly 1-unconditional, since it can happen that for $|a_n| \leq |b_n|$, $\forall n$, and $|a_{n_0}| < |b_{n_0}|$ for some n_0, that

$$\left\| \sum a_n t_n \right\| = \left\| \sum b_n t_n \right\| \ .$$

(The canonical basis for l_p is strictly 1-unconditional, for $1 \leq p < \infty$). An example: $\|t_i + t_j\| = \|t_i\| = 1$, $\forall i \neq j$.

We used the program to test what happens to the norm of a vector as we slowly 'inflate' varying coefficients, and found that:

Lemma: In Tsirelson's's space T, vectors of form

$$\theta_1 t_1 + \theta_2 t_2 + \theta_i t_i + \theta_j t_j \ , \quad (2 < i < j) \ ,$$

are extreme points of B_T, for any signs $\{\theta_1, \theta_2, \theta_i, \theta_j\}$.

Proof: By using Remark 5 (Notes and Remarks: Chapter I), and the definition of norm in T, we easily see that $\|\theta_i t_i + \theta_j t_j\| = 1$, for $i \neq j$, and signs θ_i, θ_j. Similar considerations yield that for any vectors $\sum a_n t_n$ of norm 1, the vector $\sum b_n t_n$ also has norm 1, where

$$\begin{cases} b_1 = 1 \ , \\ b_n = a_n, \text{for } n > 1 \ . \end{cases}$$

Thus no norm 1 vector $\sum a_n t_n$ with $|a_1| < 1$ is extreme. Now the permutation σ on $\mathbb{N}$ defined by

$$\begin{cases} \sigma(1) = 2 \ , \\ \sigma(2) = 1 \ , \\ \sigma(n) = n \ , \text{ for } n > 2 \end{cases}$$

induces an isometry on T (this is Theorem III.8), in the sense that $\left\| \sum a_n t_n \right\| = \left\| \sum a_{\sigma(n)} t_n \right\|$. From this we can conclude that for any $\sum a_n t_n \epsilon$ ext (B_T), $|a_1| = |a_2| = 1$. It's easy to use Remark 5 (Notes and Remarks: Chapter I) and the definition of norm to now show that vectors of form $\theta_1 t_1 + \theta_2 t_2 \theta_i t_i + \theta_j t_j$ are norm 1. For the sake of argument, assume that $\theta_1 t_1 + \theta_2 t_2 + \theta_i t_i + \theta_j t_j = \frac{1}{2}(c_1, c_2, c_3, \cdots) + \frac{1}{2}(d_1, d_2, \cdots)$, where $(c_1, c_2, \cdots)$ and $(d_1, d_2, \cdots) \in B_T$.

Then since $\|\cdot\|_{c_0}$ minorizes $\|\cdot\|_T$, $c_k = d_k = \theta_k$, for $k \in \{1, 2, i, j\}$. Now if $c := (c_1, c_2, \cdots) \neq \theta_1 t_1 + \theta_2 t_2 + \theta_i t_i + \theta_j t_j$, $\exists \ell \epsilon \{1, 2, i, j, \}$ such that $|c_\ell| \neq 0$.

202

But then $\|c\|_T \geq \|c\|_1 \geq \frac{2+|c_\ell|}{2} > 1$, where $\| \cdot \|_1$ is the 1-level norm in the ascending sequence of norms defining $\| \cdot \|_T$. But then $c \notin B_T$.   □.

A classical result of J. Lindenstrauss says that the ball of T has uncountably many extreme points, so it now becomes of interest to characterize them. This much of the above must hold: any extreme point must be ± 1 in the first two coordinates. It may be that the above program will lead to some further insight into this problem.

INDEX

Printed in the United States
by Baker & Taylor Publisher Services